Why Has ~~~ of Navy Ships Risen?

A Macroscopic Examination of the Trends in U.S. Naval Ship Costs Over the Past Several Decades

Mark V. Arena • Irv Blickstein
Obaid Younossi • Clifford A. Grammich

Prepared for the
United States Navy

NATIONAL DEFENSE
RESEARCH INSTITUTE

The research described in this report was prepared for the United States Navy. The research was conducted in the RAND National Defense Research Institute, a federally funded research and development center sponsored by the Office of the Secretary of Defense, the Joint Staff, the Unified Combatant Commands, the Department of the Navy, the Marine Corps, the defense agencies, and the defense Intelligence Community under Contract DASW01-01-C-0004.

Library of Congress Cataloging-in-Publication Data

Arena, Mark V.
    Why has the cost of Navy ships risen? : a macroscopic examination of the trends in U.S. Naval ship costs over the past several decades / Mark V. Arena, Irv Blickstein, [et al.].
      p. cm.
    "MG-484."
    Includes bibliographical references and index.
    ISBN 0-8330-3921-0 (pbk. : alk. paper)
    1. United States. Navy—Procurement. 2. Warships—United States—Costs. 3. Shipbuilding—United States—Costs. 4. Shipbuilding industry—United States—Costs. I. Blickstein, Irv, 1939– II. Title.

VC263.A799 2006
359.6'212—dc22

2006008649

Cover photo courtesy of the U.S. Navy
Photographer's Mate 3rd Class Konstandinos Goumenidis, photographer

The RAND Corporation is a nonprofit research organization providing objective analysis and effective solutions that address the challenges facing the public and private sectors around the world. RAND's publications do not necessarily reflect the opinions of its research clients and sponsors.

RAND® is a registered trademark.

Cover design by Stephen Bloodsworth

Published 2006 by the RAND Corporation
1776 Main Street, P.O. Box 2138, Santa Monica, CA 90407-2138
1200 South Hayes Street, Arlington, VA 22202-5050
4570 Fifth Avenue, Suite 600, Pittsburgh, PA 15213-1516
RAND URL: http://www.rand.org/
To order RAND documents or to obtain additional information, contact
Distribution Services: Telephone: (310) 451-7002;
Fax: (310) 451-6915; Email: order@rand.org

Preface

Recent testimony by Admiral Vernon Clark, former Chief of Naval Operations, indicated that ship costs have increased at a rate far greater than inflation. As a result, it is becoming more difficult for the Navy to afford the ships it needs in the fleet. To better understand the source of these cost increases, the RAND Corporation was asked to quantify the causes of the cost growth and suggest options to reduce it. This report documents that effort. This report should be of interest to the Navy and the Office of the Secretary of Defense, as well as congressional planners involved in ship acquisition.

This research was sponsored by the Assessment Division, Office of the Chief of Naval Operations (OPNAV N81) and conducted within the Acquisition and Technology Policy Center of the RAND National Defense Research Institute, a federally funded research and development center sponsored by the Office of the Secretary of Defense, the Joint Staff, the Unified Combatant Commands, the Department of the Navy, the Marine Corps, the defense agencies, and the defense Intelligence Community.

For more information on RAND's Acquisition and Technology Policy Center, contact the Director, Philip Antón. He can be reached by email at atpc-director@rand.org; by phone at 310.393.0411, x7798; or by mail at RAND Corporation, 1776 Main Street, P.O. Box 2138, Santa Monica, CA 90407-2138. More information about RAND is available at www.rand.org.

Contents

Figures

Tables

Summary

Over the past four decades, the growth of U.S. Navy ship costs[1] has exceeded the rate of inflation. This cost escalation concerns many in the Navy and the government. The real growth in Navy ship costs means that ships are becoming more expensive and outstripping the Navy's ability to pay for them. Given current budget constraints, the Navy is unlikely to see an increase in its shipbuilding budget. Therefore, unless some way is found to get more out of a fixed shipbuilding budget, ship cost escalation means that the size of the Navy will inevitably shrink. In fact, by some estimates, even boosting the shipbuilding budget from $10 billion annually to $12 billion would only help the Navy achieve a fleet of 260 ships by the year 2035 rather than the nearly 290 it now has (CBO, 2005).

To better understand the magnitude of ship cost escalation and its implications, the Office of the Chief of Naval Operations asked the RAND Corporation to explore several questions. These include the magnitude of cost escalation, how ship cost escalation compares with other areas of the economy and other weapon systems, the sources of cost escalation, and what might be done to reduce or minimize ship cost escalation.

---

[1] By "cost," we are technically referring to the government's "price" in the analysis sense. So, we are including not only the shipbuilder's cost and fees, but also the government's direct costs, such as government-furnished equipment and material. Although we will use the term "cost" throughout this document, formally it is more correctly "price."

Historical Cost Escalation

In the past 50 years, annual cost escalation rates for amphibious ships, surface combatants, attack submarines, and nuclear aircraft carriers have ranged from 7 to 11 percent (Table S.1). Although exceeding the rates for common inflation indexes (e.g., the Consumer Price Index [CPI]), these ship cost escalation rates have not exceeded those for other weapon systems. Over the same period of time, for example, the annual cost escalation rate for U.S. fighter aircraft was about 10 percent. Historical analyses of British Navy weapon systems also show cost escalation rates comparable to those the Navy has experienced in recent years.

Principal Sources of Cost Escalation for Navy Ships

We examined two principal groups of factors for ship cost escalation: *economy-driven* and *customer-driven*. *Economy-driven* factors are largely outside the control of the government and include elements such as wage rates and the cost of material and equipment. While some elements of these costs (e.g., health care costs reflected in burdened labor rates) have increased faster than common inflation indexes in recent decades, we found that the overall contribution of economy-driven factors to ship cost escalation was roughly comparable to that of inflation. The economy-driven factors accounted for approximately half the overall escalation. We did not observe significant improvements in labor productivity.

Table S.1
Cost Escalation Rates for Battle Force Ships, 1950–2000

Ship Type	Annual Growth Rate (%)
Amphibious ships	10.8
Surface combatants	10.7
Attack submarines	9.8
Nuclear aircraft carriers	7.4

Customer-driven factors include elements the government wants on a ship, regulations it imposes for standards and requirements in shipbuilding practices, and methods it uses to purchase ships. These customer-driven factors increase design and construction complexity, which in turn affect cost. Characteristic complexity is a measure of how changes to basic ship features (e.g., displacement, crew size, number of systems) make them more difficult to construct. Our statistical analysis found that light ship weight (LSW)[2] and power density (i.e., the ratio of power generation capacity to LSW) correlated most strongly with ship costs. Note that these relationships are associative and not necessarily causal. In other words, going to a smaller or less-power-dense ship will not always result in a lower-cost vessel. Power density, for example, was related to the number of mission systems on a ship. That is, generators do not cause the ships to be much more expensive, but the systems they are required to run do. Nonetheless, we can use these measures to gauge how the complexity of vessels has changed with time. Excepting aircraft carriers, LSW has grown by 80 to 90 percent for the ships we compare. Clearly, the Navy's desire for larger and more-complex ships has been a significant cause of ship cost escalation in recent decades.

Other standardization and requirements desired by the government have also contributed to ship costs. These include improvements in survivability, habitability, working conditions both on board and in constructing ships, and environmental regulations surrounding the construction and operation of ships. For surface combatants, it appears that the contribution of such standardization and requirements to shipbuilding cost escalation is roughly equal to that of labor, equipment, or increasing complexity of vessels. Procurement rates contributed a smaller portion to overall cost escalation.[3]

---

[2] LSW, or light displacement, is the weight of the ship (in tons) including all permanent items. It does not include variable loads such as crew, stores, and fuel.

[3] Some effects due to production rate decreases, such as increased overhead and cost escalation due to a diminished supplier base, are included with the labor and equipment categories. The influences of these factors due to rate effects could not be isolated.

To quantify the effects of the changes described above, we compared specific ship classes. In Table S.1, we calculated the overall trend for all classes of a given type. But to quantify component effects, we made pair-wise comparisons. For our example, we compare a DDG-2 authorized in FY 1961 with a DDG-51 authorized in FY 2002. The overall annual escalation rate for this comparison is slightly lower (9.1 percent vs. 10.7 percent) but of similar magnitude to that shown in Table S.1 for surface combatants. Figure S.1 shows our assessment of annual escalation rate components. The buildup of the individual factors results in an annual rate of 8.9 percent, which is very close to the observed rate. The economy-driven factors (material, labor, and equipment) account for roughly half the overall rate of increase, whereas the costumer-driven factors (complexity, standards and requirements, and procurement rate) account for the other half.

Figure S.1
Contributions of Different Factors to Shipbuilding
Cost Escalation for Surface Combatants:
DDG-2 (FY 1961) and DDG-51 (FY 2002)

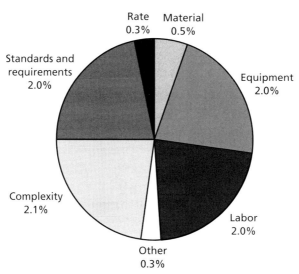

RAND *MG484-S.1*

In contrast to this 9.1 percent annual growth rate for surface combatants, the recent growth rate for the DDG-51 program shows a much more modest rate of increase. Between 1990 and 2004, the price for a DDG-51 grew, on average, by only 3.4 percent per year—a value slightly higher than the CPI over this time. Such a modest growth rate results from the fact that a relatively stable design was being produced (i.e., with no significant changes in complexity or capabilities). This observation corroborates our earlier observation that most of the growth beyond inflation is due to changes in the customer-driven factors.

Shipbuilders' Perspective on Cost Escalation

In addition to quantifying principal sources of cost escalation, we asked shipbuilders for their views on other issues contributing to increasing costs. Among the most prominently mentioned was an unstable business base. Many shipyards have a monopsony relationship with the government—that is, the government is their main, if not only, customer. At the same time, fluctuating ship orders from the Navy, with initially forecast orders typically exceeding what is ultimately purchased, discourage shipyards from making investments that could ultimately reduce the cost of ships. More importantly, an unstable business base causes fluctuations in the demand for skilled labor that are expensive and difficult to manage. The unstable business base also prevents contractors from leveraging purchases (long-term contracts) from subcontractors and suppliers that might result in more stable pricing. The shipbuilders also noted a diminished supplier base leading to single sources for many ship components (this is particularly acute in submarine manufacture). This shrinkage of the supplier base has led to higher prices and longer lead times for delivery. Finally, the unstable business base makes it difficult for the shipbuilders and suppliers to manage their workforce—that is, to hire new workers or to retain skilled workers.

Other issues contributing to cost escalation cited by the ship-builders include health care costs and equipment and material escalation due to diminished buying capacity and other market forces.

Options for Reducing Ship Costs

What might be done to reduce ship costs while supporting the fleet size the Navy desires? Unfortunately, there are no easy or simple solutions. Most approaches involve some level of compromise. Proceeding without any change will likely result in ever-diminishing procurement quantities, ultimately leading to a shrinking fleet size. To counter the increasing cost, the Navy can target some of the main factors related to escalation, such as those related to the capability and complexity of vessels. Limiting the growth in features and requirements is one approach to containing price escalation and would target roughly one-half the increase shown in Figure S.1. Indeed, where the Navy has produced a class with a relatively stable design, the cost changes have stayed in line with inflation (e.g., the recent DDG-51 experience). Another approach to contain requirements and features is to reconsider the mission orientation of ships. Rather than building large, multi-mission ships, the Navy could build smaller, mission-focused ships, thereby constraining requirements growth and reducing the cost of any single hull. A third approach to containing requirements growth is to separate the mission and weapon systems from the ship (similar to the modular approach currently being pursued with the Littoral Combat Ship). By separating the mission systems from the ship, it may be possible to reduce the total number of mission packages in the fleet (i.e., each ship does not need a complete set of mission packages).

There are areas in which the shipbuilders might be able to reduce cost. Some investment initiatives—for example, investments in lean manufacturing and shipbuilding technologies—could improve the efficiency of shipbuilding. However, some thought needs to be given to how to encourage such efficiency improvements. Traditional contracting approaches have not provided adequate incentives for the shipyards to invest. Another potential area for reduction is with indirect costs, which have grown faster than inflation. While reductions in these areas might be helpful, they only target the labor portion of the escalation (less than a quarter of the overall escalation shown in Figure S.1). Labor costs could be reduced but cannot be eliminated.

Other approaches to reduce escalation include the way we buy ships—either in program management or in acquisition strategy. For example, the government could use longer-term contracts (multiyear buys) to add some stability to the production demand. The Navy could seek to improve aspects of program management, such as reducing change orders and having better continuity of government management. The government could also consider concentrating production rather than spreading it around multiple producers. Such an approach might lead to greater efficiencies (through "learning" and overhead) but could result in the closure of some shipyards.

There are other steps that could potentially reduce the cost of building naval ships. But these items are less politically palatable, such as a rationalization of shipbuilding capacity or the involvement of foreign competition. However, Congress has been reluctant to take such steps (e.g., rejecting the "winner-take-all" competition for the DD[X] and driving a teaming arrangement for the production of the *Virginia*-class submarine).

Conclusions

The cost escalation for naval ships is nearly double the rate of consumer inflation. The growth in cost is nearly evenly split between economy-driven and customer-driven factors. The factors over which the Navy has the most control are those related to the complexity and features it desires in its ships. While the nation and the Navy understandably desire technology and capability that is continuously ahead of actual and potential competitors, this comes at a cost. We do not evaluate whether the cost is too high or low, but note only that it exists. Nevertheless, given that the pressures on shipbuilding funds will continue in the foreseeable future, the Navy may need to continue seeking ways to reduce the costs of its ships—and this will likely need to come from, in part, a limiting of the growth in requirements and features of ships. The shipbuilders can also help to reduce the cost escalation of ships through improvements in efficiency and reductions in indirect costs.

Acknowledgments

There are many individuals who contributed to this study whom we would like to thank. We would first like to thank Trip Barber of OPNAV N81 for both sponsoring this study and providing very useful input and guidance along the way. His advice and questions improved the analysis and presentation greatly. We would like to also thank CDR Todd Beltz, also of OPNAV N81, for his comments and suggestions in the generation of this report. A special acknowledgement also goes to Christopher Deegan of Naval Sea Systems Command (NAVSEA) 017. Mr. Deegan provided much of the source data for this study and made many constructive comments throughout the study. Philip Sims of NAVSEA 05 was very helpful in explaining and defining historical trends for the characteristics of Navy ships. We thank him for his insight, time, and the data he provided.

We would also like to thank the U.S. shipbuilders and their parent organizations—the General Dynamics Corporation and the Northrop Grumman Corporation—for their time and insight. Particularly, we would like to thank John Brenke (Northrop Grumman Ship Systems), Steve Ruzzo (General Dynamics Electric Boat), and Thomas Thornhill (Northrop Grumman Newport News) for coordinating our interactions with the shipbuilders and providing helpful feedback on the study.

Larrie Ferreiro of Defense Acquisition University suggested the useful comparison and provided data for the passenger ship cost analysis in Appendix E of this report. We thank him for his help and insight.

We thank Phillip Wirtz for editing and preparing the document for publication.

Finally, we would like to thank both of the reviewers of this document: John Graser, of RAND, and Daniel Nussbaum, of the Naval Post Graduate School. Their comments and suggestions have greatly improved this work.

Abbreviations

Amphib	amphibious ship
BLS	Bureau of Labor Statistics
CBO	Congressional Budget Office
CNO	Chief of Naval Operations
CFE/M	contractor-furnished equipment and material
CGT	compensated gross ton/tonnage
CPI	Consumer Price Index
CVN	nuclear aircraft carrier
DD	destroyer
DDG	guided missile destroyer
DD(X)	next-generation destroyer
DoD	Department of Defense
ECI	Employment Cost Index
FFG	guided missile frigate
FY	fiscal year
GAO	Government Accountability Office
GDP	gross domestic product

GFE	government-furnished equipment
GFE/M	government-furnished equipment and material
GRT	gross registered ton/tonnage
LCS	Littoral Combat Ship
LHA	amphibious assault ship
LPD	amphibious transport dock
LSW	light ship weight
MPF(F)	maritime prepositioning force (future)
NAVSEA	Naval Sea Systems Command
NAVSEA 05	NAVSEA Ship Design, Integration and Engineering Division
NAVSEA 017	NAVSEA Cost Engineering and Industrial Analysis Division
OPNAV N81	Office of the Chief of Naval Operations, Assessment Division
OSHA	Occupational Safety and Health Administration
POM	program objective memorandum
PPI	Producer Price Index
SCN	Shipbuilding and Conversion, Navy
SSN	nuclear attack submarine
TOA	total obligational authority
TY	then year

CHAPTER ONE

The Growth of Ship Costs

Former Chief of Naval Operations' Perspective and the Significance of the Problem

Over the past four decades, U.S. Navy ship costs have exceeded the rate of inflation. In written testimony to Congress, Admiral Vernon Clark, former Chief of Naval Operations (CNO), noted cost increases of four types of ships—nuclear attack submarines (SSNs), guided missile destroyers (DDGs), amphibious ships, and nuclear aircraft carriers (CVNs)—between 1967 and 2005 that ranged from 100 to 400 percent (Clark, 2005). The specifics for each ship type are shown in Table 1.1. Based on these values, we have calculated a real, annual growth rate (i.e., the annual increase in costs above inflation) for building these ships. It ranges from 1.8 to 4.3 percent (see Table 1.1).

This cost escalation concerns many in the Navy and the government. The real cost growth means that ships are becoming more expensive and outstripping the Navy's ability to pay for them. Given current budget constraints, including those from increasing budget deficits and costs for continued operations in Iraq, the Navy is unlikely to see an increase in its shipbuilding budget. The problems that increasing costs and fixed budgets present to the Navy are further complicated by requirements for several new ship classes such as next-generation destroyers (DD[X]s), aircraft carriers (CVN-78s), amphibious transport docks (LPD-17s), maritime prepositioning force ships (MPF[F]s), costing billions of dollars per hull.

Table 1.1
Cost Escalation of Naval Ships

Ship Class	Cost in 1967 (FY 2005 millions $)	Cost in 2005 (FY 2005 millions $)	Cost Increase (%)	Real, Annual Growth Rate (%)
Nuclear attack submarines	$484	$2,427	401	4.3
Guided missile destroyers	$515	$1,148	123	2.1
Amphibious ships	$229	$1,125	391	4.3
Nuclear aircraft carriers	$3,036	$6,065	100	1.8

SOURCE: Clark (2005).

To demonstrate how this real growth erodes the ability to buy ships and sustain a fleet, we projected how increasing ship costs causes a decrease in the number of ships per year that may be acquired. This decrease in acquisition rate, in turn, results in a smaller sustainable fleet size. Figure 1.1 shows the average number of ships per year that may be acquired as a function of time for three different budget-level assumptions. We assumed the budget levels to be fixed in real terms—that is, budgets would increase only to offset inflation. The real growth in the price of ships was the same as shown in Table 1.1. We also assumed that the Navy would buy ships in the same proportion as they exist in the current fleet—in other words, the composition of the fleet would not change. The starting cost for a given type (carrier, surface combatant, etc.) was assumed to be the same as the 2005 values in Table 1.1 (e.g., a new carrier would cost approximately $6.1 billion). For each year, we escalated that cost by the real annual growth rate shown in Table 1.1. Thus, every year each vessel becomes more expensive to acquire, while the budget remains fixed. This results in fewer ships that may be purchased.

As can be seen in Figure 1.1, there is a steady decrease in the average number of ships per year. In 2005, the average number of ships per year ranges from just over five ships for an $8 billion budget to about eight ships for a $12 billion budget. The corresponding steady-state fleet size (the largest fleet that can be sustained at the average acquisi-

Figure 1.1
Average Number of Ships Acquired per Year and Corresponding Steady-State Fleet Size Under Varying Levels of Fixed Shipbuilding Budgets

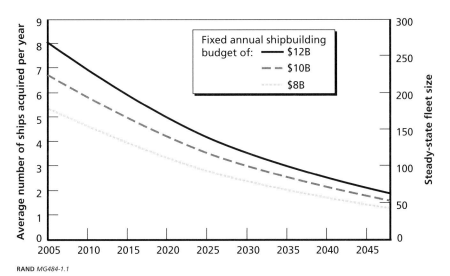

RAND *MG484-1.1*

tion rate) ranges from about 180 ships for an $8 billion annual budget to 260 ships for a $12 billion annual budget. This assumes an average ship life of 30 years except for carriers with an assumed life of 50 years. By 2025, the average number of ships per year, and their corresponding steady-state fleet sizes, is nearly halved.

Admittedly, this analysis is quite simplistic in that it does not account for a number of factors, such as a changing mix of ships, the actual forecast cost of newer proposed hulls (e.g., for the Littoral Combat Ship [LCS], LHA[R], CVN-21, or DD[X]), the fact that escalation is not uniform,[1] and the actual retirement or replacement patterns of the existing fleet. A more detailed analysis by the Congressional Budget Office (CBO)[2] estimated that, on average, 7.4 ships per year would need to be built in order to have a fleet size of 260 ships with a

---

[1] Cost escalation may not be uniform with time. For example, escalation could be minimal when producing a fixed design rather than a new design or one that has been improved. We explore this issue in greater detail in the next chapter.

[2] Bruno (2005b); CBO (2005).

corresponding annual budget of approximately $14 billion (FY 2005 dollars) (excluding the cost of nuclear refueling) for the 2006–2035 time frame.

Regardless of the top line budget number, there will be a steady decline in the real purchasing power of the Navy as ship costs escalate. All the budget curves shown in Figure 1.1 converge to lower procurement rates with time. Clearly, the Navy will not be able to sustain a fleet of nearly 300 ships at these acquisition rates, and the problem will only become more difficult with time.

Ship Cost Escalation and Complexity

Cost escalation for weapon systems and the difficulties that result from it is not a new problem, nor one limited to naval ships. Two decades ago, Norman Augustine, having demonstrated that the cost of an aircraft increased by a factor of four every ten years, famously quipped, "In the year 2054, the entire defense budget will purchase just one aircraft. This aircraft will have to be shared by the Air Force and the Navy three and one half days each per week except for leap year, when it will be made available to the Marines for the extra day" (Augustine, 1986). Augustine observed that aircraft unit cost was more closely related to the passage of time than modifications in speed, weight, or technical specifications. This "law" has, over time, been considered to apply to other military weapon systems.

Navy ships also have sources of high and increasing costs that are unique compared with other weapon systems and commercial ships. Much of their high cost lies in the fact that the design and construction of naval ships is one of the more complicated tasks of weapon system engineering and manufacturing that a country can undertake. Naval shipbuilding requires both heavy manufacturing and high-tech systems integration, including a complex integration of communication, control, weapons, and sensors that must work together as a coherent system. These components, or subsystems, are a mix of various technologies, including electronics, mechanical systems, and software. These technologies, particularly for weapon systems, are state of the art and may still be undergoing development when a program begins.

Beyond their direct military mission, naval ships must perform so-called hotel functions associated with housing and feeding hundreds of sailors who staff the ship. Warships also need to provide for the health of the crew and thus require medical facilities. All these capabilities must be sustained for several months at sea, requiring a significant amount of equipment and provision storage. These non-mission capabilities of warships make them unique compared with other military assets, such as tanks and aircraft.

Given the size and complexity of warships, manufacturing them requires substantial design, engineering, management, testing, and production resources. The workforce at a typical naval shipyard numbers in the thousands and includes many engineering specialties (e.g., electrical, mechanical, naval architecture). Modern naval ships are designed using sophisticated, three-dimensional computer-aided design tools, requiring a highly skilled and educated workforce. Production requires such diverse skills and trades as electricians, welders, and pipe fitters. Testing complex systems on ships requires commissioning and test specialists to verify functionality; for some skills—for example, those performed by nuclear-qualified welders and commissioning engineers—it might take years to become proficient.

U.S. naval ship production predominantly serves one customer: the U.S. government. The products are fully tailored (i.e., customized) for the mission of the vessel. In other words, few existing designs can serve as a basis for modification as is usually done in commercial shipbuilding. Naval ship production rates are low compared with those for commercial ships and production varies from three years to more than a decade. Production is allocated between producers when there is more than one shipyard capable of producing a class.

Despite these differences in product, market, and manufacturing for naval and commercial ships, naval shipbuilding is often compared with other industries in the consumer economy, with observers frequently commenting on the lack of benefits from a highly competitive market with multiple buyers and sellers and the attendant efficiencies gained through high-volume production. The expectation of such comparisons is that naval shipbuilding retains many of the dynamics of the commercial shipbuilding industry. This report addresses, in part, the validity of such comparisons and what may be learned from them by identifying specific areas in which naval shipbuilding costs have exceeded those for commercial industries and some of the reasons for this greater escalation.

Study Objectives and Overview

The escalation in ship costs and its implications recently led the Office of the Chief of Naval Operations (OPNAV) to ask the RAND Corporation to explore several questions related to ship cost escalation, including:

- What has been the magnitude of cost escalation for Navy ships?
- How does this cost escalation compare with other areas in the economy and with other weapon systems?
- What are the sources of the cost escalation for Navy ships?
- Can this escalation be reduced or minimized?

Approach

Our approach is a "top-down" analysis that highlights and explores the issues related to ship cost[3] escalation and what, if anything, can be done to mitigate it. This work takes a "macro-level" approach, examining overall industrial and technological trends and their correlation with ship cost. We analyze ship cost and economic data to define the trends and factors related to cost escalation, including how technical, performance, capability, requirement, and other variables have changed and might influence cost escalation.

Our core concern, as noted, is cost escalation. We use this term to describe the general changes, typically for a similar item or quantity, in cost between periods of time. We distinguish between *cost escalation*

---

[3] By "cost," we are technically referring to the government's "price" in the analysis sense. So, we are including not only the shipbuilder's cost and fees, but also the government's direct costs, such as government-furnished equipment and material. Although we will use the term "cost" throughout this document, formally it is more correctly "price."

and *cost growth*. *Cost growth* is traditionally defined as the difference between actual and estimated costs. We are not concerned with evaluating these; rather, we are studying how the actual cost for an item changes as time passes.

Cost escalation can be measured by cost increase. Cost increase is the percentage change in cost between time periods. Algebraically, it is

$$\frac{Cost_2}{Cost_1} - 1, \tag{1.1}$$

where

- $Cost_2$ is the cost at time period 2
- $Cost_1$ is the cost at time period 1.

If, for example, $Cost_2$ is \$5 and $Cost_1$ is \$4, then the cost increase is 0.25, or 25 percent.[4]

Because we examine *cost increases* over varying periods of time, we calculate *annual growth rates* to normalize cost increases to a common baseline. Algebraically, we define annual cost growth as

$$rate = \sqrt[(Year_2 - Year_1)]{\frac{Cost_2}{Cost_1}} - 1, \tag{1.2}$$

where

- $Cost_2$ is the cost at $Year_2$
- $Cost_1$ is the cost at $Year_1$.

That is, the annual growth rate is a compound function in which year-to-year increases accumulate. If, for example, $Cost_2$ is \$5 and $Year_2$ is 2004, and $Cost_1$ is \$4 and $Year_1$ is 1998, then the resulting annual

----

[4] Mathematically, this is (5/4) − 1, or 1.25 − 1, or 0.25, or 25 percent.

growth rate for cost may be calculated as 3.8 percent.[5] "Real" annual growth rates are calculated by using a constant dollar basis (one corrected for inflation).

To organize the analysis and simplify presentation, we split the cost growth factors we examine into two broad categories. The first category comprises *economy-driven* factors, inputs to ship cost that are largely outside the government's control.[6] These may include worker wages and benefit costs, labor productivity, indirect labor costs, and material and commodity equipment costs.[7]

The second category comprises *customer-driven* factors, centering on the nature of the product and how it is acquired. These may include such characteristics as size, speed, power generation, stealth, survivability, habitability, and mission and armament systems. In general, a more complex and larger ship will cost more than a smaller and simpler one. Customer-driven factors also include those related to acquisition strategy—such as the number of ships purchased, the timing of purchases, and the number of producers receiving work—and their effects on government costs, as well as government policies directly targeted to shipbuilding, such as worker compensation and environmental regulations. As stated previously, the "customer" is both the Navy and the federal government.

Alternatively, one may analyze the sources of cost escalation through a formal engineering analysis entailing a series of detailed technical evaluations of specific ship classes. In other words, one could

---

[5] Mathematically, the terms in this example are $Year_2 - Year_1 = 2004 - 1998 = 6$ and $Cost_2/Cost_1 = 5/4 = 1.25$. The sixth root of 1.25 is approximately 1.0379; subtracting 1 from this gives an annual growth rate of 0.0379, or approximately 3.8 percent.

[6] Of course, the federal government can and does influence the general economy through fiscal and monetary policy. Such policy, however, is not targeted to the approximately $10 billion naval shipbuilding industry, which remains a relatively small portion of the approximately $12 trillion economy. We therefore view government fiscal and monetary policy as a part of overall economic conditions and as not malleable for naval shipbuilding purposes.

[7] The Navy can also influence some of these costs through its choice of material (e.g., grade of steel) for ships or its setting of indirect rates through its procurement practices and purchasing patterns. The Navy does not, however, directly influence the commodity price of such materials or the rates for such labor. The shipyards and, ultimately, the Navy must pay the market price for such items. Hence, we classify these, too, as not customer driven.

explore the specific technical differences between systems (e.g., mission, weapons, and ship), requirements, and standards. One might compare the acquisition cost and performance differences for two ships, such as cost differences for the Aegis SPY-1A and SPY-1D radar systems, or compare how costs have evolved over time for painting and preservation standards of tanks and voids. Resources for this study and client interests, however, dictated that we pursue the top-down approach, to present results both in a timely fashion and in a way that encompasses as many relevant broad topics as possible. Other organizations, such as the Naval Sea Systems Command's (NAVSEA's) Ship Design, Integration and Engineering (Code 05), and shipbuilders have analyzed some of the more detailed issues. We draw upon their work to supplement and support the high-level analysis we have conducted.

As the data allow, we will examine trends from the 1950s through today. This time frame was selected to be consistent with the CNO's analysis. However, we do explore whether the time frame affects our conclusions. Appendix D evaluates the time trends from the 1990s to today.

Sources of Data

We use ship cost data provided by the NAVSEA Cost Engineering and Industrial Analysis Division (Code 017). Primary data were the final "end unit costs" for various Navy ships (by hull) going back approximately five decades. These end unit costs represent the total cost for a ship, including government-furnished equipment (GFE) and advanced procurement funds, but not the related research and development monies. The values are based on the final budget submissions for each hull and are the best long-term, final cost data that were available. The ship and unit costs were also broken down into the standard budget P-5 Exhibit[8] format (planning costs, basic construction and conversion, propulsion equipment, ordnance, electronics, etc.). NAVSEA 017 also provided average engineering and production hours for each class.

---

[8] The P-5 Exhibit is a budget breakdown of the costs for a major weapon system. Such exhibits are part of the annual "Justification of Estimates: Shipbuilding and Conversion, Navy" as part of the Navy's budget submission.

To explore how the physical characteristics of Navy ships have evolved in recent decades, we also analyze data on light ship weight (LSW), power generation, shaft horsepower, and crew size. These data were obtained from multiple sources, including the Assessment Division, Office of the Chief of Naval Operations (OPNAV N81); NAVSEA 05; NAVSEA 017; the Naval Vessel Register; and shipbuilders. General Dynamics also provided data on the cost changes due to other, less-measurable features and manufacturing changes, such as survivability improvements and the effect of Occupational Safety and Health Administration (OSHA) regulations.

Finally, to compare cost escalation for naval shipbuilding to that in other industries and the overall economy, we use data compiled by the Bureau of Labor Statistics (BLS).[9]

Report Organization

Our analysis is presented in the next three chapters. Chapter Two analyzes the historical cost escalation for ships and compares it to other weapon systems as well as other sectors of the economy. Chapter Three explores the sources of cost escalation by dividing them into two broad classes: economy-driven and customer-driven. Chapter Four discusses the issue of cost escalation from the perspective of shipbuilders. Following this analysis, we discuss, in Chapter Five, potential approaches to mitigate or control cost escalation. We present our conclusions in Chapter Six. Appendixes A and B provide details on the multivariate regressions model for ship cost. Appendix C lists the questions we asked industry. Appendix D presents analysis on recent naval ship cost growth (1990–2004). Appendix E analyzes the cost growth of passenger ship as a further point of comparison.

---

[9] Available at its Web site, http://www.bls.gov.

CHAPTER TWO

Historical Cost Escalation for Ships

As we have noted, cost escalation for ships has outpaced general infla-tion in recent decades. This cost escalation points to several questions we analyze in this chapter, including:

- Is this cost escalation prevalent in other time frames?
- Is the escalation linear with time, or does it have some other func-tional form?
- How does this trend compare with varying sectors of the civilian economy or other defense sectors?
- How does this escalation compare with other weapon systems, such as aircraft and missiles?

Cost Escalation for Navy Ships

The form of the cost escalation trend over time can provide some insight into its underlying reasons. For example, if the increase is con-tinuous and steady, then escalation may be due to a systemic issue such as the increase of shipbuilding input cost (e.g., greater costs for labor or materials). A sustained increase might also be due to increased costs for a continuously improving product, in which each ship of a given class is continually upgraded or uses the most advanced technology avail-able at the time of production. A more periodic increase may indicate an escalation due to periodic changes in technologies, requirements, or capabilities when new designs are introduced.

Surface Combatant Example

We illustrate in Figure 2.1 the form of cost escalation over time for three surface combatant types—DDGs, destroyers (DDs), and guided missile frigates (FFGs)—produced between 1950 and 2000.[1] Surface combatants are the single largest group of ships in the Navy's ship battle forces—comprising nearly one in three such vessels—and therefore it is the group of ships for which we have the most data. Each point in the figure represents the cost and fiscal year budgeted for a single hull. The solid line represents the best exponential fit through the DDG data; because the end unit cost axis scale is logarithmic, it appears as a straight line on the lognormal plot. This line shows an annual growth rate in ship costs (as measured in then-year [TY] dollars) for these ships.

The shifts in DD and DDG costs tend to follow a "stairstep" pattern. This reflects the way the Navy develops and produces ships—in discrete classes in which a relatively fixed design is produced for a

Figure 2.1
Cost Escalation for Selected Surface Combatants

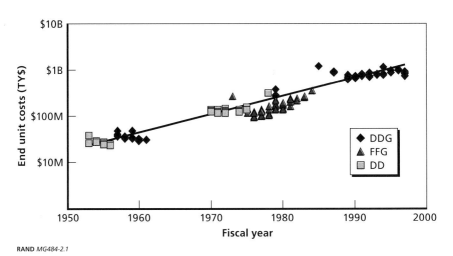

RAND *MG484-2.1*

---

[1] For simplicity, we restrict this example to surface combatants. The other combatant types have similar histories, but the surface combatants are more numerous. We also restrict the surface combatants to DDs, DDGs, and FFGs so that the trends are more discernable.

period of time. The onboard systems for a class may improve over time, but the basic design of the ship does not change.[2] When a new class is introduced, the costs "step" to a new plateau. "Steps" also appear when there are significant changes in capabilities between classes or through deliberate evolution to a new flight. Where ship content and capabilities remain stable over time, cost growth is more modest. This behavior can clearly be seen in the DDG cost history for the DDG-2s, DDG-993s, and the DDG-51s. The DDG-2s were produced from the mid-1950s through the early 1960s. Notice that the costs initially start higher then slowly decrease over the production run, reaching a "plateau." There was a step up for the DDG-993s in the late 1970s. The DDG-51s stepped up to a higher cost level, but also show this plateau after the initial few hulls.

These trends suggest that the causes of cost escalation for DDs and DDGs may be driven by evolutions in capabilities of ships rather than increases in basic shipbuilding costs such as those for labor or materials. If labor and material cost increases were causing cost escalation for these ships, then we would see an upward trend within and between classes.

The FFG data, which are solely for the FFG-7 class, do not manifest a similar "stairstep" trend. Nevertheless, its cost escalation still appears attributable to an evolution of its capabilities and roles. During the course of its production, this class evolved from a single-role ship for antisubmarine warfare to a multi-mission surface combatant. The class was initially envisioned as the "low" end of the high-low mix concept (Federation of American Scientists, undated). This high-low mix concept was seen as a way to maintain or increase fleet size and control acquisition costs by purchasing some simpler, mission-focused ships (low end of mix), not as versatile as the multi-role ships, but that would augment the capabilities of the high end, highly capable multi-role ships and fill a niche the more expensive ships

---

[2] There are variants produced within a class of ships. For example, the DDG-51 class has evolved through three distinctive flights (I, II, and IIA). These changes tend to be more evolutionary improvements (e.g., upgraded systems) with the occasional addition of more capability (e.g., a hangar for helicopters).

did not. The low-end ships tended to be smaller and have fewer systems and were therefore less expensive to produce. Analyzing the specific components of costs shows how these increased capabilities contributed to cost escalation.

Figure 2.2 shows how basic ship costs (i.e., shipbuilder costs), electronics (GFE) costs, and ordnance (GFE) costs changed for this class of ships between 1973 and 1984, the years in which construction of the FFG-7 was authorized. Each of these costs is presented as a proportion of costs in 1973 (e.g., a value of 150 percent in 1975 indicates that the unit cost for the component grew by 50 percent [in then-year dollars] above its 1973 cost). Basic and ordnance costs remained relatively stable in the time period shown by this chart (even decreasing in then-year dollars), while the cost of GFE electronics increased more than fivefold. The increased electronics cost resulted from the additional capability that was added to the class for its expanded roles.

Figure 2.2
P-5 Component Escalation for the FFG-7 Class

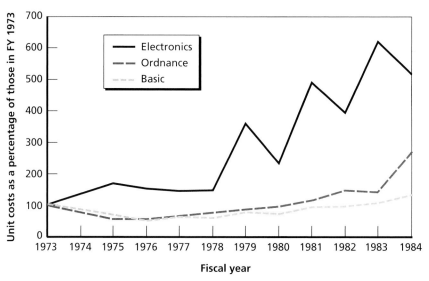

Comparing Cost Escalation Among Ships

How does cost escalation for surface combatants compare to that for other ships? We present data on this in Table 2.1, showing annual growth rates of costs, unadjusted for inflation. The rates are based on the end cost data provided by NAVSEA 017 over the years 1950 to 2000.

Escalation rates for these ships in the past 50 years have varied between 7 and 11 percent per annum—that is, the rate for surface combatants does not differ greatly from that of other ships. Excluding the carrier data, cost escalation for surface combatants appears to be nearly identical to that for the other ships. One possible explanation for the lower escalation rate for nuclear carriers is that the data for nuclear aircraft carriers span a more limited range of time, from FY 1958 to FY 1995, and nearly all these carriers are for *Nimitz*-class hulls. As a result, the carrier data, unlike data for the other battle force ships, do not reflect the cost escalation that occurs when new classes of ships are developed.

Cost Escalation for Other Weapon Systems

How does cost escalation for Navy ships compare with that for other weapon systems? To examine whether there is a unique pattern or trend to cost escalation for Navy ships, we analyzed the cost escalation for U.S. fighter aircraft from 1950 to 2000 (see Figure 2.3). Each symbol

Table 2.1
Cost Escalation Rates for Battle Force Ships, 1950–2000

Ship Type	Annual Growth Rate (%)
Amphibious ships	10.8
Surface combatants	10.7
Attack submarines	9.8
Nuclear aircraft carriers	7.4

Figure 2.3
Fighter Aircraft Cost Escalation, 1950–2000

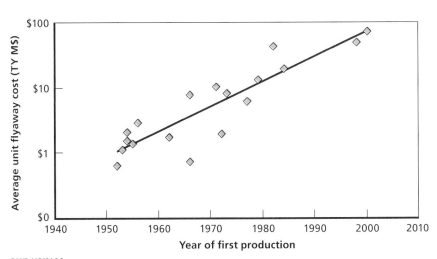

shows the average unit flyaway cost for a series of different aircraft as a function of the year production started for the series.

Because we have only the average unit cost for each aircraft type, we cannot discern whether aircraft follow a stairstep pattern similar to that evident across classes of Navy ships. Nevertheless, we can calculate an overall escalation rate for fighter aircraft, indicated by the solid line representing the best exponential fit to the data. This indicates a 9.3 percent annual growth rate in the cost for fighter aircraft, which is very similar to the annual growth rates for naval ships indicated in Table 2.1.

Cost escalation is also evident in naval ships and weapon systems of other nations. In Figure 2.4, we present real, annual escalation for a number of UK weapon systems based on the work of Philip Pugh (1986). Because these data, unlike those in Table 2.1, have been adjusted for inflation, they should be comparatively lower than those we have shown for U.S. ships and other weapon systems. In sum, cost escalation appears to be a systemic problem for all weapon systems, and not one limited to construction of U.S. Navy ships.

Figure 2.4
Cost Escalation for UK Weapon Systems

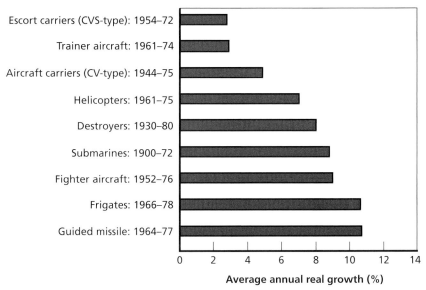

SOURCE: Pugh (1986).

Cost Escalation in Other Sectors of the Economy

How does naval ship and weapon system cost escalation compare to other sectors of the economy? To examine this, we compare it with the Department of Defense (DoD) Comptroller's historical deflator, the gross domestic product (GDP) deflator, and with the BLS Consumer Price Index (CPI).

DoD Deflator

The DoD Comptroller publishes a deflator for the purposes of forecasting and normalizing expenditures to a common basis, based on the total obligational authority (TOA) deflators for procurement reported in annual national defense budget estimates (see Office of the Under Secretary of Defense [Comptroller], 2004, p. 46). This set of indexes forms the de facto inflation adjustment for weapon systems.

The average annual growth rate for the TOA procurement deflator between 1965 and 2004 was approximately 4.6 percent. This rate is considerably less than the escalation rates shown for Navy ships over a similar time frame. The DoD deflator, however, adjusts for price differences of *identical* commodities over time, not those that have been improved or whose capabilities have expanded, as naval ships' have. It is meant to adjust for changes in wage rates, commodity prices, and other variables similar to those in the "economy-driven" factors we consider and will explore further in Chapter Three. Recently, the DoD deflator has been much lower than other measures of inflation (see Appendix D for details).

GDP Deflator

The GDP deflator (sometimes called the GDP implicit price deflator) is a measure of inflation produced by the U.S. Department of Commerce, Bureau of Economic Analysis. This inflation index measures price changes in the overall economy. Unlike some other indexes (e.g., the CPI), it is not based on a fixed basket of goods. Because it is based on the total value of all goods and services produced, it is considered a more representative measure of inflation for an economy. Between 1965 and 2004, the GDP index (Chained Price Index) showed an annual compound rate of 4.1 percent.

Consumer Price Index

Perhaps the best-known measure of changes in consumer prices is the CPI, published by the Bureau of Labor Statistics. This index measures price changes to a sample of typical consumer goods. Between 1965 and 2004, the CPI grew at an annual compound rate of 4.7 percent. As Figure 2.5 shows, growth in the CPI is similar to that in the DoD deflator, although the DoD deflator grew more rapidly than the CPI in the 1970s, and more slowly in the 1980s and 1990s—as did the GDP deflator. The patterns of growth for these indexes, featuring exponential growth in the 1970s but more linear growth since then, also differ from that for ship cost escalation, which has been exponential throughout this time frame.

Figure 2.5
CPI, DoD TOA Procurement Deflator, and GDP Deflator Trends Since 1965

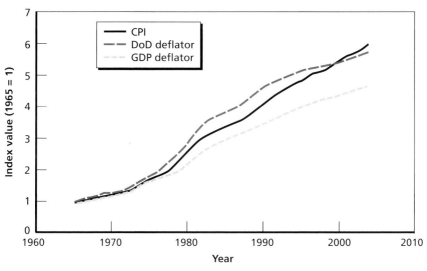

RAND *MG484-2.5*

The CPI comprises several components, some of which have increased more rapidly than others. Table 2.2 summarizes annual growth rates for a select set of components for as many years as available since 1965 (some components, such as college tuition, are not reported as far back as 1965).

CPI components exhibit a wide range of growth rates, although none are as high as naval ship escalation rates. The highest, and the closest to naval ship rates, is that for college tuition, at 8.0 percent. Medical care, at 6.6 percent, is the second highest on the list.

Summary

The long-term cost escalation for naval ships, with annual growth rates ranging from 7 to 11 percent, is much higher than rates of inflation measured by the CPI, the DoD procurement deflator, or the GDP

Table 2.2
Annual Growth Rate of Selected CPI Components

CPI Component	Annual Growth Rate (%)
Apparel	2.4
Private transportation (new and used cars, repairs, fuel, etc.)	4.2
Food and beverage	4.6
Gasoline	4.9
Shelter	5.5
Medical care	6.6
College tuition	8.0
Naval ships	7.0–11.0

implicit price deflator. In general, the actual escalation is more periodic, increasing significantly with the introduction of a new ship class. Within a class, the growth tends to be quite modest. Although exceeding that for other industries, naval ship cost escalation is comparable to that of other weapon systems. Therefore, cost escalation is not just a problem for naval ships but also for weapon systems generally. In the next chapter, we will explore the sources contributing to this escalation for naval ships.

CHAPTER THREE

Sources of Cost Escalation for Navy Ships

Types of Cost Escalation

There are several noteworthy potential sources of cost escalation for naval ships. Perhaps the most obvious is the cost for basic inputs, such as labor, material, and equipment, used in shipbuilding. If the costs for these inputs increase, the additional costs are largely passed to the government. The indirect costs for labor and manufacturing change as well. The government has enacted legislation to protect the health of workers, requirements that result in additional costs. Certain worker benefits, particularly those regarding health care, have become more expensive. As a result, if two identical ships were to be built at two different points in time, we would expect differences in cost resulting from these factors, although some increases may be offset by productivity improvements made by suppliers or shipbuilders in repeating production processes.

The Navy does not, however, build identical ships over time. Threats change, technology improves, and operational doctrine evolves. As a result, the Navy changes what it buys to meet warfighting needs. Over time, the Navy may purchase ships that are more complex, survivable, and capable; such changes to the ships have cost implications.

In this chapter, we explore how these sources of escalation have affected naval ship costs. As noted in our introductory chapter, we divide these factors into two categories: economy-driven and customer-driven.

Economy-driven factors are those largely outside the control of the government. Examples include wage rates and costs of material and equipment. Although the government can exert some influence on these costs, economic policy and worker health and safety protection are rarely targeted to the shipbuilding industry; rather, these factors are more systemic and should affect all shipbuilding programs uniformly.

Customer-driven factors relate to how a ship is built and acquired, as well as the features of that ship. Ship characteristics such as size, speed, power generation, stealth, survivability, habitability, and mission and armament systems are among customer-driven factors influencing ship costs. In general, a larger and more complex ship will cost more than a smaller and simpler one. The effect of these factors will depend greatly on specific systems and programs. Two contemporaneous programs, for example, may experience different cost escalation relative to a previous class based on changes in design, regulations, and acquisition strategy.

Comparing Ship Costs Across Time

In comparing ship costs across time, we present two types of data: (1) data on factors that are generic for all ship classes, such as labor cost escalation for the naval shipyards, and (2) other data factors specific to particular ship classes, such as changes in LSW and production rates. In some of our multivariate modeling, we also include variables for ships (e.g., auxiliaries) not of primary concern here so as to construct more accurate equations.

When assessing and validating escalation rates and their contributing factors within given groups of ships, we make comparisons between specific classes. Such comparisons allow us to address changes in both economy-driven and customer-driven factors in ship cost escalation. Making comparisons across specific classes also allows greater transparency of the contribution of individual components to ship cost escalation. The ship types and pairs (and fiscal years) of comparison classes we chose for our cost escalation analysis within classes include the following:

- Surface combatant: DDG-2 (FY 1961) to DDG-51 (FY 2002)

- Attack submarine: SSN-667 (FY 1965) to SSN-777 (FY 2002)
- Amphibious ship: LPD-1 (FY 1959) to LPD-18 (FY 1999)
- Aircraft carrier: CVN-68 (FY 1967) to CVN-76 (FY 1995).

The attack submarine, amphibious ship, and aircraft carrier pairings represent specific hull number–to–hull number comparisons as well as comparisons across classes. For example, in comparing the costs of SSN-667 to SSN-777, we compare costs of the *Sturgeon* class to those of the *Virginia* class. The surface combatant pairings represent comparisons of average costs for classes; in particular, we compare average costs for *Charles Adams*–class destroyers acquired in FY 1961 to those of *Arleigh Burke*–class destroyers acquired in FY 2002. In selecting comparison pairs, we selected those that represented as closely as possible the time frames of the examples featured in the CNO's recent testimony to Congress (see Table 1.1 in Chapter One).

In the sections that follow, we will show the contributions of various factors to the overall escalation rate. In practice, the calculations are slightly more involved, since growth factors (using Equation 1.1) were initially calculated for each factor, combined (either by adding or multiplying, as appropriate), and then converted to an annual growth rate (using Equation 1.2). For example, labor, material, and equipment growth factors were added together to determine the growth factor for the economy-driven factors. The growth rate for the economy-driven factors was then converted into an annual growth rate. We prorated the individual labor, material, and equipment component contributions to this annual growth rate according to the magnitude of the growth factor. For simplicity, we present annual growth rates for the components we consider.

We turn next to analysis of economy-driven and customer-driven factors for each of these pairs of comparisons.

Economy-Driven Factors

We first review labor as a proportion of ship costs for each of the four types of ships we examine, then how labor costs have escalated in comparison to other costs. We subsequently do the same for material and equipment costs.

Labor

Labor costs represent a substantial portion of the total procurement for naval ships. Table 3.1 summarizes the typical labor percentage of total cost for the four ship types. These labor percentages represent general averages provided by NAVSEA 017 for recent ship classes.[1] Labor costs are fully burdened, including direct and overhead costs for all types of labor (e.g., engineering, support, manufacturing). These labor costs are for the shipbuilder's contribution only. We do not have data to examine labor costs at the supplier or subcontractor level.

Labor constitutes between 32 and 51 percent of the construction costs for the ships we analyzed. (The remaining cost is split between material and equipment, both contractor- and government-furnished.) Quantifying the contribution of labor to cost escalation is therefore important to understanding total cost escalation for ships.

In Figure 3.1, we display the average annual growth rate for shipyard labor for the "Big Six" shipyards,[2] calculated from NAVSEA 017 data from 1977 to 2005.[3] We present both unburdened and burdened

Table 3.1
Labor as Percentage of End Cost by Ship Type

Ship Type	Labor % of End Cost
Nuclear aircraft carrier	51
Amphibious ship	47
Attack submarine	39
Surface combatant	32

[1] Ideally, we would analyze data on how the labor, material, and equipment percentages have changed over time, but such data were not available.

[2] These shipyards are those currently involved in U.S. naval ship production: Bath Iron Works, Electric Boat Corporation, NASSCO (National Steel and Shipbuilding Company, a division of General Dynamics), Northrop Grumman Newport News, and Northrop Grumman Ship Systems (Ingalls and Avondale).

[3] No information was available before 1977.

Figure 3.1
Shipyard Labor Rate Escalation, 1977–2005

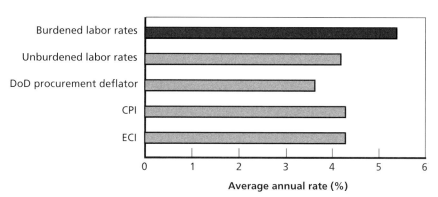

RAND *MG484-3.1*

rates. Unburdened labor rates are the direct charge rates for all labor, exclusive of overhead, benefits, and corporate charges. This rate corresponds to typical worker pay rates. The burdened rate is the cost the Navy pays for labor and includes all indirect factors (e.g., benefits, capital depreciation, maintenance). Also shown in the figure are the CPI, the DoD procurement deflator, and the Employment Cost Index (ECI) for durable goods manufacture. The ECI, generated by the BLS, reflects changes in wages and benefits for workers in particular industries.

Not surprisingly, the direct rate grew at a rate comparable to that of the CPI, which is typically used as a proxy for changes in cost of living, and to the ECI, which shows broad trends in wages. That is, direct pay for shipyard workers kept pace with consumer inflation and wages in other manufacturing industries. The average burdened rate, however, grew faster than the rates of both the CPI and the DoD deflator. The shipyards have attributed this greater increase in the burdened rate to the rise in health care costs (see Table 2.2, which shows the costs for medical care to be among the most rapidly growing among CPI components), increased disability costs, and a declining business base (with less work requiring that fixed indirect costs be spread over fewer hours of labor). Therefore, some of the cost escalation for ships that has outstripped inflation can be attributed to increases in labor costs.

As labor costs increase, industries seek productivity gains to offset them. To determine whether there are any productivity gains that have offset labor costs, we examined whether there has been a reduction in the number of direct labor hours per ton of ship. NAVSEA 017 provided average hours per hull for approximately 40 ship classes between 1970 and 1995. For each class, we have a single value, not several, so we were not able to correct for any "learning" effect experienced in constructing a particular class. Figure 3.2 shows the trend slopes (linear regression) for two ship types—surface combatants and auxiliaries—in hours of labor per LSW in tons, by year in which a class was first authorized. There were insufficient data to present trends for other ship types. For

Figure 3.2
Class Average Light Ship Hours per Ton by First Fiscal Year of Construction for Class

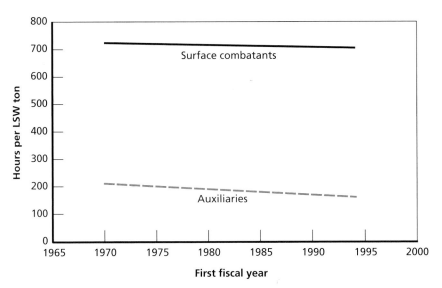

both surface combatants and auxiliaries, there was a slight decrease in hours per ton, but these changes were not statistically significant.

Because the measure we use for productivity is imprecise, it may mask changes that have occurred as ships have become more complex or as contractors have outsourced more construction tasks. Another limitation to these data is that they present only average values per class—meaning that any recent improvements, particularly for current classes that have been in production for several years, would be masked. Even so, a recent study of the shipbuilding industry has noted efficiency gains at the shipyards (Office of the Deputy Under Secretary of Defense [Industrial Policy], 2005).

A better way to measure productivity changes would be to examine the hours to complete a specific task where the content of the task remains fixed over time. Other analysts have used compensated gross tonnage (CGT) rather than LSW to examine shipbuilding productivity.[4] Unfortunately, we do not have data at the levels of detail necessary to conduct such analyses. Shipbuilders do point to specific investments and process changes (e.g., lean manufacturing, new facilities) that they claim have led to productivity gains. Because we cannot quantify such productivity improvements, we will not correct the labor factor for ship escalation. This means that we have likely overstated the impact of labor on cost escalation.

Another area related to labor costs is profit levels and fees. Typically, the shipbuilders earn a fee based on their labor costs (although the details are contract specific). One possible explanation for cost escalation could be that fee levels have increased. Unfortunately, we do not have cost data to analyze the fee time trend. However, fees are typically a small percentage of the labor costs (generally less than 10 percent) and so changes in fee levels would have, at most, a modest effect on ship cost escalation.

––––––––––––––

[4] The CGT approach is a method used in commercial shipbuilding to adjust gross weight of ship or vessel based on a complexity factor. Ships of low complexity have low factors, and ones of higher complexity have higher factors. Thus, productivity measures of hours per CGT attempt to correct for complexity. See Craggs et al. (2003a,b) and Lamb (2003) for a discussion of the use of CGT for naval shipbuilding productivity.

Material and Equipment

Material and equipment form the remainder of costs for a given naval ship. These items can include commodities varying from steel plate to complex weapon systems. More generally, material includes basic items used in shipbuilding such as steel, paint, electrical cable, and insulation. Equipment includes major manufactured items such as systems (e.g., for navigation, weapons, or command and control), machinery (e.g., elevators, pumps, air conditioning units), or electrical distribution (e.g., switchgear, circuit boards, transformers).

NAVSEA 017 provided the percentage of total ship cost for government-furnished equipment and material (GFE/M) and contractor-furnished equipment and material (CFE/M). Based on these data, we derived percentages of material and equipment in total ship construction cost by assuming that GFE/M is entirely equipment and that CFE/M is split evenly between material and equipment (see Schank, Pung, et al., 2005). Given these assumptions, we present in Table 3.2 the approximate percentages of material and equipment that comprise construction costs for each type of ship. Equipment costs range from 35 to 57 percent of construction costs for the ship types we analyzed, while material costs range from 11 to 17 percent.

To understand how material and equipment costs contribute to naval ship cost escalation, we analyzed escalation rates for several components of the Producer Price Index (PPI). The PPI measures change in selling prices received by domestic producers of goods and services. The PPI components we selected as being representative of the material and equipment used in shipbuilding were the following:

Table 3.2
Equipment and Material as Percentage of Construction Costs by Ship Type

Ship Type	Equipment (%)	Material (%)
Surface combatant	57	11
Attack submarine	46	16
Amphibious ship	37	17
Nuclear aircraft carrier	35	14

1. Electrical machinery and equipment (BLS series ID: WPU117)
2. Electronic components and accessories (BLS series ID: WPU1178)
3. Switchgear, switchboard, relays, etc., equipment (BLS series: ID WPU1175)
4. Mechanical power transmission equipment (BLS series ID: WPU1145)
5. General purpose machinery and equipment (BLS series ID: WPU114)
6. Steel mill products (BLS series ID: WPU1017).

We stress that this list does not represent all components involved in shipbuilding but is meant only to be a sample of such items. Other components, such as aeronautical, nautical, and navigational instruments (WPU118501) might also be selected, although these particular PPI data do not exist prior to 1985 and hence would have been of limited usefulness in making comparisons to ships built in the 1960s.

Figure 3.3 displays the average annual growth rate between 1965 and 2004 for the PPI components we selected. For comparison, we also show the average growth rate for the CPI and for the DoD procurement deflator during this time. The annual growth rate in cost for the components we selected was at or below those for the CPI and the DoD deflator.

To determine aggregate rates of escalation for material and equipment, we must average PPI components together. Because we do not know the appropriate weightings for components (i.e., the contribution of each to the total cost of a ship), we make some simplifying assumptions for this analysis. We assume that material cost escalation follows that for steel mill products. We ignore electronic components and accessories, because it is not clear whether they belong in equipment or material categories. We assume that equipment cost escalation follows the average for the other four PPI components we analyze. With these assumptions, we calculate the material and equipment annual escalation rates shown in Table 3.3. Both rates are below the CPI and DoD deflators for the 1965–2004 time frame.

Figure 3.3
Material and Equipment Cost Escalation, 1965–2004

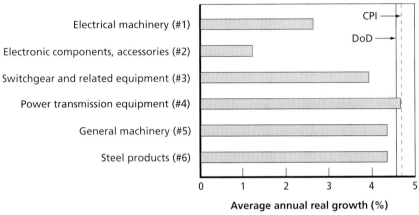

Table 3.3
Material and Equipment Annual Escalation Rates, 1965–2004

Commodity	Annual Escalation Rate (%)
Material	4.4
Equipment	3.9

Similar to labor, shipbuilders typically charge a material overhead for purchased material and equipment. The escalation rates in Table 3.3 are exclusive of this overhead charge. We do not have data to examine how the material overhead rate has changed with time.

Summary of Economy-Driven Factors

Labor costs have increased at a rate greater than inflation, while material and equipment costs have increased somewhat less so. How much of the overall cost escalation do these factors explain? In Table 3.4, we list the contributions to escalation by labor, material, and equip-

Table 3.4
Contributions to Annual Cost Escalation by Labor, Material, and Equipment

Ship Type	Economy-Driven Factors				Actual Rate (%)
	Labor (%)	Equipment (%)	Material (%)	Total (%)	
Surface combatants	2.0	2.0	0.5	4.5	9.1
Attack submarines	2.4	1.7	0.5	4.6	9.9
Amphibious ships	2.9	1.3	0.7	4.8	8.2
Aircraft carriers	2.8	1.7	0.7	5.2	7.1

ment for the four ship types we analyze (numbers may not sum to total due to rounding). Each of these factor contributions was determined by examining how the representative indexes changed over the time period in question and then weighting each by that factor's percentage of the total cost (shown in Tables 3.1 and 3.2). For example, if labor costs represented 30 percent of the total ship costs and the escalation rate for labor was 7 percent over that time frame, the contribution for labor would be 2.1 percent ($0.3 \times 0.07 = 0.021$). We also list in the table total escalation rates for each pair type of ship classes that we consider. (These rates differ from those shown in Table 2.1 because they cover a different time period.)

Labor comprises the largest of the three economy-driven factors, with the greatest increase over the time period analyzed. Still, the total contribution from all three factors ranges only from 4.5 to 5.2 percent—or about one-third to one-half less than the actual escalation rates. In other words, the economy-driven factors can only explain about half the total cost escalation seen for ships and, by themselves, represent a cost escalation roughly equal to that of inflation for the time period.

What else may be contributing to cost escalation? In the next section, we examine the contribution of customer-driven factors to the difference between economy-driven factors and the actual rate of escalation.

Customer-Driven Factors

What and how the customer chooses to buy can influence the cost of a naval ship. By "customer," we mean more than the Navy; other elements of the government, including Congress and the Office of the Secretary of Defense, also influence the cost. Ship cost varies by content and size. A larger, more complex ship will be more expensive than a smaller, simpler one. A nuclear aircraft carrier, for example, costs much more than a coastal patrol vessel. A highly capable ship with many mission systems, such as an Aegis destroyer, is more expensive than a refueling or resupply ship. Purchasing a ship can be considered analogous to purchasing an automobile. Smaller cars of the same class (e.g., family sedan, compact car) tend to cost less than larger ones, and adding options increases the price of the car.

Other actions by the customer can also influence ship price. For example, the federal government sometimes directs production of a class (e.g., DDG-51 and SSN-774) to multiple producers. The Navy has recently argued that the new DD(X) class could be produced more effectively with a sole-source contract (Bruno, 2005a); Congress, however, does not want to pursue this strategy (Capaccio, 2005), a decision that may indirectly affect the cost of the ship. Other government actions, including general regulations for the environment and occupational safety and health, as well as those regulations targeted to the shipbuilding industry, such as placing workers under the Longshoreman's Act, can influence production costs as well. Such regulations, of course, may have broader benefits to society that outweigh their contributions to ship costs, but the fact remains that they still affect ship costs and should be considered in discussion of the sources of cost escalation.

The most germane point about customer-driven factors of ship cost escalation is not that ships are complex and expensive or that government regulations increase cost. Rather, it is whether these factors have *changed* over time. That is, have ships become more complex over time? Have regulations and requirements changed, and contributed to increasing costs, over time? We consider these questions below.

Characteristic Complexity

Complexity of weapon systems is very difficult to define quantitatively, particularly across different systems.[5] For our purposes, we define complexity as the difficulty and level of effort required to design, manufacture, integrate, and outfit a ship. Some ships (e.g., a patrol vessel) require less effort to build than others (e.g., a nuclear submarine). To understand how the complexity of warships has evolved, we need quantitative measures that reflect how ships today differ from those produced a few decades ago. Because we lack data on ship costs and hours at levels at which we can examine changes in specific ship systems or areas, we use the basic characteristics of a ship (e.g., displacement, crew size, number of systems) as proxies for complexity, or, as we call it, "characteristic complexity." The advantage of selecting common ship characteristics is the resulting ability to look at trends across ship types by an objective measure (or one not based on a subjective evaluation or scale).

In selecting ship characteristics to analyze, we chose among those that are readily available and could be determined across decades. Some measures that may be applicable to modern weapon systems, such as software lines of code, are not applicable for this study given that few, if any, ships constructed before 1965 had advanced computing systems. Similarly, we cannot use measures such as maximum depth that apply only to certain ship types (e.g., submarines).

Table 3.5 lists the characteristic factors we considered and how they are proxies for complexity. The Navy (OPNAV N81) provided data for LSW, shaft horsepower, and electrical power generation. Crew size was obtained from the Naval Vessel Register.[6] Counts for both number of armament and mission systems were based on ship specifications listed by GlobalSecurity.org and the Federation of American Scientists[7] for each ship class and flight.

---

[5] Some examples of measuring the changes in complexity for complex systems have been conducted in the automobile industry. See, for example, Alexander and Mitchell (1985); Fisher, Griliches, and Kaysen (1962); and Triplett (1969).

[6] Online at http://www.nvr.navy.mil.

[7] Online at http://www.fas.org.

Table 3.5
Ship Characteristics to Measure Ship Complexity

Factor	Definition	Relation to Complexity and Cost
Light ship weight (also called light displacement)	The weight of the ship (in tons) for all permanent items. It does not include variable loads such as crew, stores, and fuel.	LSW is a proxy for size. Larger ships should cost more than smaller ships, other things being equal (same functionality, class, etc.).
Shaft horsepower (SHP)	The maximum, total output power generated from the engines for propulsion (in horsepower).	SHP is a measure of the size of the propulsion plant. The greater the SHP, the larger or greater the number of engines and the more complex shaft and gear system.
Electrical power generation	Electrical power generation (in megawatts) for hotel and systems on the ship. It does not include electrical power generation for the propulsion plant.	Electrical power generation is a proxy for the number and complexity of the systems on board the ship. It is not necessarily about the cost of generators, but rather the items on the other end of the wire. Admittedly, this is somewhat of an imperfect measure because systems have become more power efficient over the past few decades (e.g., going from vacuum tubes to transistors and other solid-state devices).

Table 3.5—continued

Factor	Definition	Relation to Complexity and Cost
Crew size	The number of personnel accommodations on the ship (both enlisted and officer) in terms of a headcount.	A larger crew size means a bigger ship (more space required) as well as more supporting facilities (e.g., messes, berthing areas, heads). A larger crew could also be related to having a greater variety of mission systems on board, since most require specialists to operate. However, reducing crew size dramatically can also increase cost. Typically, manpower reductions are done through increased automation, which increases procurement costs.
Number of mission systems	A count of the number of mission systems (e.g., communications, sensors, electronic support measures/electronic countermeasures, navigation equipment, and fire control).	The number of systems should correlate directly to the cost (more systems cost more) and the level of integration difficulty.
Number of armament systems	A count of the number of armament systems on board (e.g., missiles, torpedoes, guns).	The number of armament systems should correspond to the cost (more systems cost more).

To determine the effect of complexity on ship cost, we used multivariate regression analysis of ship costs and ship characteristics. Our approach is similar to that for developing a cost estimating relationship, except our purpose is to understand the relative differences in cost for changes in characteristics, not to forecast future cost. NAVSEA 017 provided end unit costs by hull for 37 ship classes; these classes are listed in Appendix A. Note that the data sample has too much variation in ship type to be useful as a cost estimating relationship—for example, it includes a mix of nuclear and nonnuclear ships.

Our analysis proceeded in two stages. First, we used cost improvement analysis to determine the hypothetical first unit cost in FY 2005 dollars and the improvement slope by class, flight (where applicable), and shipbuilder. That is, we fit each unique class, flight, and shipbuilder subset of the data to the functional form of the following equation:

$$C_n = C_1 \times n^{\ln(slope)/\ln(2)}, \tag{3.1}$$

where

- C_1 is the hypothetical first unit cost
- n is the unit number (e.g., $n = 2$ corresponds to the second hull produced)
- C_n is the cost for the nth unit produced
- *slope* is the cost improvement slope in decimal form.

The evaluation yielded 56 unique observations for C_1 and slope by class, flight, and shipyard.

Second, we regressed the various characteristics on C_1 using stepwise multiple regressions. The best fit for the regression was for a log-log formulation. We also included an independent term for the ln(*slope*) in the regressions. The coefficient of the ln(*slope*) term in the initial regressions indicated that the best convergence point for the regression was the ninth unit for each observation; put broadly, though not precisely, the ninth unit is generally the point at which a full "learning effect" for productivity has occurred.[8] We then redid the stepwise regressions using $\ln C_9$ as the dependent variable. The results for the regression are summarized as

$$\ln C_9 = 0.95 \ln LSW + 0.94 PowDen - 1.3 Aux + 0.29 Sub + 11.5, \qquad (3.2)$$

where

- $\ln C_9$ is the natural log of cost for the hypothetical ninth unit (fit by a cost improvement curve) in thousands of FY 2005 dollars
- $\ln LSW$ is the natural log of the LSW in tons
- $\ln PowDen$ is the natural log of the power density (i.e., electrical power generation capacity in kW divided by LSW tons)
- Aux is a binary term indicating whether a ship is an auxiliary vessel
- Sub is a binary term indicating whether a ship is a submarine.

(Appendix B provides full regression diagnostics. We included auxiliaries in this equation in order to have more degrees of freedom.)

The ship characteristic terms in the equation are LSW and power density. We choose power density rather than power generation capacity because LSW and power generation capacity are highly correlated—that is, larger ships tend to have a greater power generation capacity. LSW and power density, however, are not correlated and hence fulfill the requirement of regression analysis that independent terms not be correlated, lest they lead to misleading coefficients and estimates of significance (multicollinearity) (Berenson and Levine, 1996).

Power density may also be a better measure of complexity than power generation capacity because it is indicative of how many systems are put on a ship of a given size. Figure 3.4 shows how power density for surface combatants has evolved over the past three decades, with an approximately 40 percent increase in average power density from 1970 to 2000. This increase in power density might partially explain

---

[8] For a more complete discussion of this approach, see Younossi et al. (2003).

Figure 3.4
Power Density Trend for Surface Combatants, 1970–2000

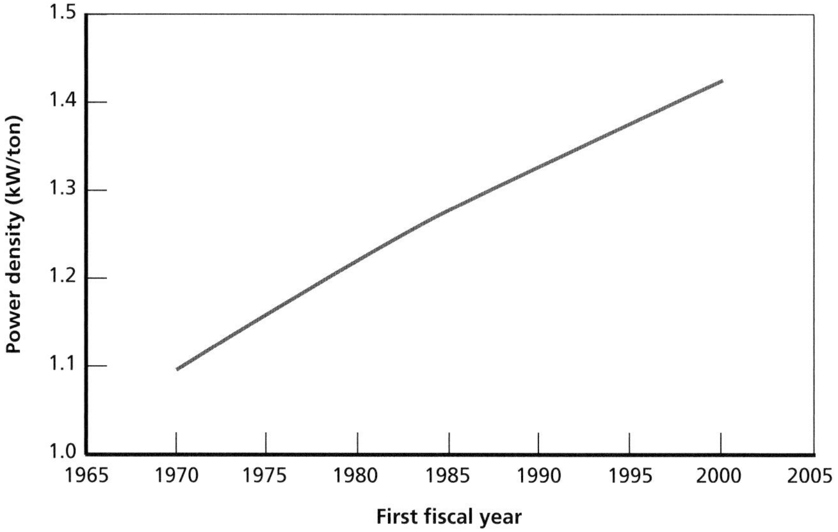

why we observed no significant change in shipbuilding productivity as measured by hours per ton: Navy ships are becoming more complex and difficult to integrate.

It is noteworthy that the number of mission systems, armament systems, and crew size were not significant terms in the regression. Rather, these terms were correlated with other, more significant terms. The number of mission systems was highly correlated with power density. The numbers of armament systems and crew size were highly correlated with LSW. LSW and power density terms in Equation 3.2 therefore implicitly control for differences in systems and crew size.

The important information relative to ship cost escalation from Equation 3.2 is the magnitude of the coefficients for lnLSW and ln$PowDen$. These two coefficients are close to unity. Because this equation is logarithmic, the coefficients indicate that doubling the size of the ship will approximately double its cost, as will doubling its power density.

With this information, we can now translate complexity changes to ship cost escalation. Table 3.6 lists the changes in LSW and power density for the ship comparisons listed earlier. Note that in all but the carrier example, the LSW of the ship increased more than 80 percent. Increases in power density were more variable.

Other Ship Features[9]

There are other aspects to complexity not included in basic ship characteristics. Many of these are represented by the "invisible" factors Philip Sims (1983) described as measures of warship effectiveness. While the "traditional" factors are similar to those we used to measure and discuss characteristic complexity, the "invisible" factors are more difficult to measure. We present a simplified list of both in Table 3.7.

Among the "invisible" mission capability factors, naval ships have clearly improved in several of them, in turn suggesting several possible sources of cost escalation. Survivability—that is, the ability of a ship to protect its crew and function despite damage—has improved through improved design, greater system redundancy, and better material choices. Habitability has also improved, with sailors now having better living and working conditions than they did 30 years ago (Sims,

Table 3.6
Contributions to Annual Escalation Rate by Characteristic Complexity

Ship Type	Change in LSW (%)	Change in Power Density (%)	Annual Escalation Rate Due to Characteristic Complexity (%)
Surface combatants	+81	+88	2.1
Attack submarines	+82	+19	1.6
Amphibious ships	+90	+32	1.7
Aircraft carriers	−1	0[a]	0.0

[a] Because we do not have power-generation data by hull, we have assumed that this has not changed for the *Nimitz* class.

---

[9] Much of the qualitative discussion in this section is based on a March 5, 2005, memo from NAVSEA 05 to RAND researchers outlining the differences between the DDG-2 and DDG-51 classes and on discussions with naval shipbuilders.

Table 3.7
Mission Capability Factors

Traditional	Invisible
Displacement	Survivability
Visible weapons	Reaction time
Visible armor	Reliability and maintainability
Speed	Endurance
Acquisition cost	Pollution control
	Seakeeping
	Habitability
	Radar and noise signature
	Sensor range
	Life-cycle cost

2004). Figure 3.5 shows the trend in the average living space per sailor (i.e., personnel-associated volume divided by the number of accommodations) for surface combatants following World War II (Kehoe, 1976). Each point represents the design values for a ship class plotted at the year the class was first operational. These data indicate that living space more than tripled in the three decades following the war.

The quality and sophistication of weapon systems carried on board has also improved. Figure 3.6 shows the steady progression of weapon systems for surface combatants. With each new class or flight, weapon systems on combatants have improved with each newer version. For example, ships are progressively able to track more targets at a given time. Their detection range has also increased. The ability to strike inland has increased in terms of range and accuracy. The relative positioning for each ship class in the figure is not meant to be exact, but rather a notional representation of how weapons on ships have become more complex over time.

Other changes have also improved ship performance in various operational dimensions. Propulsion plants now use gas turbines rather than steam engines. Higher-strength steel alloys have helped to reduce weight. Better hull designs have reduced radar cross-section. Better machinery and improved isolation systems have reduced radiated noise.

Figure 3.5
Average Living Space per Sailor on Surface Combatants, 1945–1975

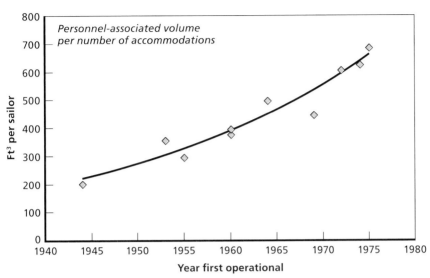

The "invisible" factor of pollution control is evident in environmental regulations that have changed the way ships treat and deal with waste. Plastic wastes are now collected and compressed and returned to land for disposal rather than disposed at sea. Degradable garbage is now shredded or pulped before disposal. Waste and contaminated water is now either collected for shore treatment or cleaned before release. All these changes require facilities that were not on ships in earlier decades.

Working conditions for both shipyard workers and ship crew members have improved. Environmental, health, and safety regulations have reduced worker exposure to toxic materials, hazardous substances, and noisy work conditions. Shipyards now control and treat production waste to a greater degree than they did a few decades ago. The working environment for the ship's crew is noise controlled and, in most cases, climate controlled.

Figure 3.6
Increasing Complexity of Weapon Systems for Surface Combatants

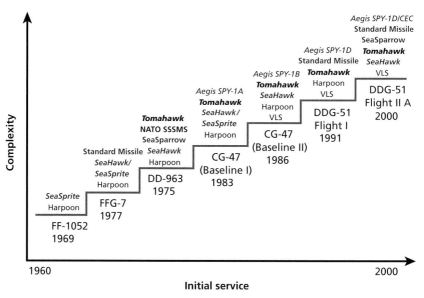

RAND *MG484-3.6*

All these improvements in capabilities, safety and survivability, manufacturing standards, and environmental policies have come with added cost. In many cases, these changes have provided significant benefit to the Navy and the nation. Nevertheless, the narrower topic of concern for this study is: How much have these changes increased the cost of naval ships? A thorough quantitative analysis of this question would require a much greater level of detailed cost data than is available to us.

To better gauge how these changes may affect shipbuilding costs generally, we rely on a 1998 analysis by the Electric Boat Division of the General Dynamics Corporation to explore how submarine costs have been driven by changes in several "invisible" areas. A condensed list of these areas includes the following:

- Technical—for example, improvements to performance, service life, maintainability, ship control, and habitability
- Stealth—reductions in noise and other signatures

- Weapons—for example, increase in the number and complexity of the systems carried, improved target tracking and acquisition, reduction in weapon handling and response times
- Quality assurance and oversight—for example, increased oversight by multiple government groups, material control and tracing, vendor certification, and external audit agencies
- Survivability—for example, improved shock resistance, fire suppression, and defect control
- Regulatory and worker environment—for example, OSHA regulations, environmental protection (air and water discharges, soil contamination, and waste handling), and workers' compensation laws.

The Electric Boat study examines how the construction cost changed during the evolution of five classes of attack submarines: SSNs -637, -688, -688i, -21, and -774. The change in cost from SSN-637 to SSN-774 due to the above factors was nearly fourfold. Although the exact study results remain proprietary, for purposes of this analysis we examined factors listed above, removed those (e.g., weapons load) addressed by characteristic complexity, and determined annual growth rates in cost for submarines and for all other ships due to standards, regulations, and requirements complexity. Table 3.8 lists the two rates, with that for submarines higher than that for all other ships. The difference between these two rates is mainly attributable to the greater stealth requirements for submarines.

Table 3.8
Cost Escalation Due to Standards, Regulations, and Requirements

Vessel Type	Annual Escalation Rate (%)
Submarines	2.6
All other ships	2.0

Procurement Practices

There are several factors the customer controls in procurement practices that influence costs. We explore two of these: the rate at which the government purchases ships and the effect of involving multiple producers.

Procurement Rates. The number of ships procured in a fiscal year can influence the price of an individual ship. Relatively higher quantities can lead to economies of scale in manufacture and purchasing. For example, the timing of units can be scheduled better when there is more certainty regarding quantity and delivery dates. Investments to improve efficiency are easier to justify when there is a more certain production run and there are more units over which to offset a fixed investment. Better price stability might also result because shipbuilders are able to enter into longer-term contracts with their suppliers. Such benefits are typical of multiyear procurements.

To evaluate the effect of procurement rate on cost escalation, we re-examined the data on individual hull cost used to determine the factors for characteristic complexity. We fit these data to a form of the cost improvement equation that included a rate term

$$C_n = C_1 \times n^{\ln(b)/\ln(2)} \times r^{\ln(c)/\ln(2)}, \qquad (3.3)$$

where

- C_n is the cost for the nth unit
- C_1 is the cost for the first unit
- n is the unit number
- b is the unit cost improvement slope (see Fisher, 1970)
- r is the number of units procured in a fiscal year
- c is the rate slope.

Within our data sample used for the analysis earlier (Equation 3.1), there were very few programs in which the procurement rate varied sufficiently to determine the unknown parameters by regression analysis (i.e., the procurement rate had sufficient variability over the procurement time frame, and the procurement time frame was greater than three years). Those programs that did have sufficient variability were AKR-310, CG-47, CVN-68, DD-963, DDG-51, FFG-7, SSBN-726, TAKR-200, SSN-668, AGOS-1, TAO-187, TATF-166, LHD-1, MCM-1, and MHC-51. As in our earlier cost improvement analysis, we determined independent terms for each unique combination of shipbuilder, class, and flight (so there were more observations than the simple count of the number of classes). Requirements for regression analysis further restricted our analysis; because highly correlated independent terms yield inaccurate coefficients, we omitted observations where n, the unit number, and r, the number of units procured in a single year, had a correlation coefficient greater than 0.6 for each shipyard/class pair. Table 3.9 presents our results for c, the rate slope.

Table 3.9 indicates that the rate slope has a mean value of 0.90. This can be interpreted as the rate the unit cost for a given unit changes as a function of the procurement rate. If the procurement rate doubles, the unit cost decreases by 10 percent. If the procurement rate is halved, the unit cost increases by 10 percent. Table 3.10 shows the cost impact for the change in the average number of units acquired per year for each of the comparison ship types.

Table 3.9
Summary Statistics for Rate Slope

Statistic	Value
Number of observations	16
Mean	0.90
Standard deviation	0.075
Minimum	0.74
Maximum	1.08

Table 3.10
Annual Escalation Rate Due to Procurement Rate

Ship Type	Average Procurement Rate Change	Annual Escalation Rate Due to Procurement Rate (%)
Surface combatants	5.8 to 3.0	0.3
Attack submarines	2.3 to 1.0	0.3
Amphibious ships	2.2 to 0.63	0.5
Aircraft carriers	N/A	0.0

Change in procurement rates has a relatively small effect on the overall escalation rate. For example, although the average annual procurement rate for amphibious ships produced between the LPD-1 and LPD-17 classes decreased from 2.2 ships per year to 0.63 ships per year, we calculate that the annual growth rate in cost resulting from this decreased procurement rate was only 0.5 percent. (Because we examine only one class of aircraft carriers, we cannot calculate the effects of changing procurement rates across classes.) This analysis does not directly measure the effect that the overall loss of business base would have on increasing indirect rates for the shipbuilder (the effect of having fewer hours across which to charge indirect labor costs was addressed in our section on labor rates). Shipbuilders have also indicated that the loss of business base has led to increased supplier cost escalation, a result of more items being produced by only a single vendor; the effect of this shrinking of the supplier base has not been included in our calculations. Such an analysis would not apply to a situation in which there are significant gaps in production, which might occur if shipbuilding rates became quite low and/or programs were stretched to fit within the budget.

Multiple Producers. For certain programs, multiple shipbuilders are involved in production, either producing whole ships or producing and integrating parts of ships. Multiple producers are often justified as a means to preserve portions of the industrial base and to add some competition to a program; the reality is that using multiple producers can make a program more politically palatable. To discern the cost effects of using multiple producers, we examined unit costs for the ninth unit (C_9) and cost improvement slopes (unit slopes) for statistical differences for ships with multiple producers. We found no statistically significant evidence that using multiple producers leads to lower unit costs or steeper improvement slopes as one might expect with competition. In fact, although we did not analyze total program costs, RAND analyses of other weapon systems suggest that using multiple producers can make a program more expensive, because multiple producers may not make it as far down the learning curve as a single one will during a constant production run (see Birkler, Graser, et al., 2001).

Summary of Customer-Driven Factors

How do customer-driven factors compare with each other in their contribution to ship cost escalation? We summarize the effects of each customer-driven factor for the annual growth rate for each ship comparison pair in Table 3.11. This table indicates that all customer-driven factors account for about half the actual growth rate shown in the far right column.

Table 3.11
Contributions to Annual Escalation Rate by Customer-Driven Factors

Ship Type	Customer-Driven Factors				
	Characteristic Complexity (%)	Standards, Regulations, and Requirements Complexity (%)	Procurement Rates (%)	Total (%)	Actual Rate (%)
Surface combatants	2.1	2.0	0.3	4.4	9.1
Attack submarines	1.6	2.6	0.3	4.5	9.9
Amphibious ships	1.7	2.0	0.5	4.2	8.2
Aircraft carriers	0.0	2.0	0.0	2.0	7.1

Total Contribution of Factors

How do economy-driven and customer-driven factors compare with each other? Table 3.12 summarizes the contribution of each set of factors to the annual growth rate of construction costs for each ship comparison pair we studied. About half the cost escalation for each is attributable to economy-driven factors, and about half is attributable to customer-driven factors. We also note the contribution of "learning" over the course of production, shown in the Cost Improvement Correction column. This correction adjusts for the fact that, in some cases, we are comparing the ship classes at different points in a production run. For example, the SSN-637 (FY 1965) was the sixth unit, and the SSN-774 (FY 2002) was the third.[10] These corrections were made only for attack submarines and amphibious ships; no corrections were made for surface combatants because both units compared were well past the ninth units produced, and none was made for carriers because

Table 3.12
Contributions to Annual Escalation Rate by Customer-Driven Factors

Ship Type	Economy-Driven Factors (%)	Customer-Driven Factors (%)	Cost Improvement Correction (%)	Total (%)	Actual (%)
Surface combatants	4.5	4.4	0.0	8.9	9.1
Attack submarines	4.6	4.5	0.6	9.7	9.9
Amphibious ships	4.8	4.2	−0.4	8.6	8.2
Aircraft carriers	5.2	2.0	0.0	7.2	7.1

---

[10] This comparison is more complex, since the SSN-637 class used a propulsion plant design produced numerous times before. Furthermore, the production of the SSN-774 class is a complex split between the two producing shipyards, so the actual unit number is something less than four.

carrier production is too infrequent to expect a learning effect resulting from repeated production.

Although the components we examine do not sum precisely to the total annual cost escalation evident and depicted in the rightmost column, the results for this high-level analysis, with all sums within 0.4 percent of the total shown, indicate that we have likely identified with reasonable accuracy the contribution of differing factors to shipbuilding cost escalation. In the next two chapters, we turn first to a more detailed analysis of shipbuilders' perspectives on these issues, then to some measures that might be undertaken in response to them.

CHAPTER FOUR

Industry Views on Ship Cost Escalation

In addition to quantifying the contribution of major factors related to cost escalation for shipbuilding, RAND researchers visited some shipbuilders to solicit their perspective on related issues. These contractors provided presentations and follow-up material in response to a series of questions, shown in Appendix C, that we provided in advance of our visits. The discussions were frank and were conducted without attribution. We provide an overview of issues common across all firms without evaluating the validity of our respondents' claims. We also draw on industry views reported in the press and other publicly available sources. We focus here on issues not addressed in previous sections of this report. We divide these issues into four general categories: unstable business bases, shrinking vendor bases, workforce issues, and increasing government regulations.

Unstable Business Bases

Until the early 1960s, the U.S. Navy owned a number of shipyards and constructed some of its own ships, but this is no longer the case. In recent years, there has been a monopsony relationship between the government and some shipbuilders. That is, for some shipbuilders, the government has been the main, if not the only, customer. The business base for these shipbuilders is closely tied to the Navy's demands.

For a brief period in the 1990s, some of these shipbuilders ventured, unsuccessfully, into commercial shipbuilding. These firms committed substantial capital in new tooling and modern facilities to compete in the global commercial shipbuilding market. For example, Newport News Shipbuilding agreed to construct nine oil tankers for a Greek firm but had to cancel the contract after delivering only five tankers, because it found itself unable to compete with foreign shipbuilders (Ma, 2005). Other firms also ventured into building cruise ships and tankers, but had similar problems. These endeavors proved to be unwise financial decisions and further tied the industry's fate to government decisions.

Despite being the industry's mainstay, government procurements have not been steady, and future forecasts for them are uncertain. The Navy demand for ships has fluctuated over time, with ultimate procurement falling short of original projections. For example, the Navy's initial order for 24 DD(X) ships was cut to eight in 2004 and further reduced to five to six ships in 2005 (O'Rourke, 2005). The Navy's plan to buy 12 LPD-17 amphibious transport ships has been cut to less than nine ships (Matthews, 2005). The Navy also envisioned buying 30 *Seawolf*-class submarines, but ultimately purchased only three (Lerman, 2005). The fluctuating requirement is exacerbated by Navy cost estimating and budgeting practices that have resulted in unrealistic funding of programs, as well as a lack of contingency for potential cost growth (GAO, 2005).

In Figure 4.1, we present broader historical comparisons of actual and projected procurement. The solid, heavy line shows actual DoD procurement expenditures, while the lines with points show projected procurement spending reported in program objective memorandums (POMs) for each year since 1984. In nearly all cases, POM projections were more optimistic than actual expenditures. Until the mid-1990s, DoD had spent less on procurement than it initially said it would.

Shipbuilders contend that these fluctuating demands and the resulting business base instability have increased unit costs (through labor workforce fluctuations) and given them less incentive to make the investments necessary to modernize their facilities and otherwise seek to improve productivity and reduce overhead costs.

Figure 4.1
Actual DoD Spending Compared with POM Projections

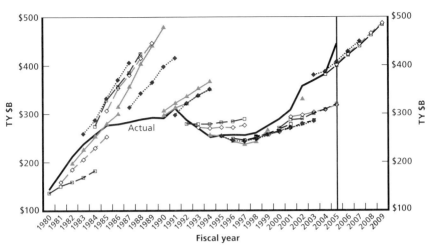

RAND *MG484-4.1*

Shrinking Vendor Bases

Shipbuilders contend that the government's reduced demand for ships has not only affected them but also vendors that provide subcontracted parts and materials. This decreased demand has led many industry contractors to consolidate through mergers and acquisitions. Many key suppliers have also left the business entirely or were unable to survive financially. Some shipbuilders, as part of their cost-cutting and stream-lining measures, have reduced the number of suppliers in an attempt to form strategic business alliances and gain leverage from larger buys; they have also reduced their procurement and quality assurance staffs.

U.S. producers of raw materials and specialized ship parts that meet military specification requirements have also decreased substantially. As a result, the lead time necessary to acquire some critical materials and parts has increased substantially; the lead time necessary for acquiring steel plate, main condensers, main motors, and nuclear pipes has almost doubled over the past three years.[1] A recent Government Accountability Office (GAO) report cites, in addition to underbudgeting for material expenditures, the limited supplier base for highly specialized and unique materials and reduced competition as other contributors to the material costs growth (GAO, 2005).

Shipbuilders rely on sole-source suppliers now more than ever before. More than 75 percent of *Virginia*-class suppliers are sole sources. Reduced rates of procurement and lack of multiyear commitments can force shipbuilders to pay premium prices for hard-to-find products and to meet contract delivery schedules.

Shipbuilding at both prime and subcontractor levels is a very captal-intensive industry. Requirements for large capital investment exacerbated by unclear business prospects can make market entry prohibitively expensive and risky for new firms.

In sum, shipbuilders feel the shrinking and unstable requirements of the government have caused many of them to consolidate business and reduce supplier bases for critical materials and parts, and that many vendors of critical items, not seeing a future in naval shipbuilding, left the business completely, further reducing the supply base. Consequently, the small sole-source vendor base for many components and raw materials has led to increased prices and longer lead times for some critical commodities.

Workforce Issues

A shipbuilding executive recently summarized the workforce issues he faces by noting that most of his company's workforce is either less than 35 years of age and has less than five years of experience or more than 45 years of age with more than 20 years of experience (Petters, 2005). That is, the industry faces issues related to both an aging workforce and "green labor." Other shipbuilders have raised similar concerns.

---

[1] Significant material cost escalation is a recent phenomenon in the shipbuilding industry. Since our empirical analysis in the previous section is based on older data (up to 2002 for actual ship costs), it does not reflect this trend. The effect of higher material costs is being seen on ships now under construction.

As a consequence of this age structure in its workforce, the industry faces the prospect of a wave of retirements in about a decade with relatively inexperienced workers to take the place of these older workers. Compounding these problems is the fact that recruiting for the skills required in the shipbuilding industry is challenging, and therefore vacancies are not easy to fill. Shipbuilding is tough work, and the requirements for labor are driven by unstable demand. Young workers may look to high-tech or other service industries as an alternative career path with better compensation and more favorable working conditions. One shipbuilder claimed that a skilled electrician would make much more money and have better working conditions in residential construction than in shipbuilding.

The requirement for new workers is exacerbated by fluctuations in ship production and alterations in the Navy's acquisition strategy. This changing environment makes it nearly impossible to make a sound business case for the recruitment of new workers (Dur, 2005). Training new workers with the right skills is time consuming and expensive. The unstable business base also makes challenging retention of highly skilled workers and those with unique skills. As production quantities fluctuate and the time between deliveries is extended, the industry must keep workers on the payroll with little work for them to do. The alternative is to let workers go and hope to rehire them at some later time. Either way, such workforce and workload variability is costly. Ultimately, these costs are reflected in the price for ships.

High health care costs and frequent use of health care by an aging workforce can further increase overhead costs. A GAO report on defense acquisition corroborates these claims and highlights medical and pension benefits as contributors to the increased shipbuilders' overhead rates (GAO, 2005). However, this phenomenon is not unique to the shipbuilding industry; it is generally considered to be one of the major sources of cost increases for many U.S. products and services.[2]

In sum, the shipbuilding industry is facing several issues that may unfavorably affect its workforce. The industry is confronting labor

_____

[2] One shipbuilder mentioned that its indirect rates had increased by about 20 percent in the past decade as a result of medical and workers' compensation costs.

issues related to an aging workforce; difficulties in attracting younger, skilled workers who may have other employment opportunities; and resulting overhead costs that ultimately may contribute to shipbuilding cost escalation.

Increasing Government Regulations

Shipbuilders we interviewed generally agree that government contractual, regulatory, and statutory requirements have increased substantially in recent decades. Contractual requirements include preparation of various reports to provide the government insights into technical, schedule, and financial performance of each contract. Among the requirements that shipbuilders claim affect their ability to pursue the most cost-effective ways of obtaining materials and products are initiatives for small business preferences, requirements to purchase American-made components, and other similar regulations.

Regulations such as the Clean Air Act require new-generation low-volatile paint and lead processing methods that involve substantial research and development and capital investment. The Clean Water Act requires water treatment capabilities and reporting of any pollutant discharge. One contractor noted that it now has to collect rainwater runoff from its site and treat it before it enters the water supply. Although workplace safety requirements, managed by OSHA, have not changed substantially since the 1970s, other recent regulations related to asbestos, lead, methylene chloride, and forklift use in the workplace have had a direct impact on the shipbuilding industry. The Sarbanes-Oxley Act of 2002, designed to strengthen corporate accountability, requires that shipbuilders (and firms in other industries) conduct regular audits, recordkeeping, reporting, and certification. Such requirements have also affected the cost of managing shipbuilding programs.

Quantifying the effects of these regulations is difficult. Shipbuilders we interviewed provided little specific empirical evidence of the effects of these regulations other than to note their sheer number. Some industry managers claimed that the cost impact was reflected in direct and indirect cost increases and could not be separately accounted.

Summary

Altogether, shipbuilders identify unstable business bases, shrinking vendor bases, workforce issues, and increasing government regulations among the contributors to cost escalation, but the cost impacts are difficult to quantify. Fluctuating demands and a decreasing business base have increased unit costs while giving shipbuilders little incentive to make investments that would increase productivity and reduce unit costs. Fluctuating demands and a decreasing base have also reduced their supply base and made it difficult for them to recruit skilled workers to the industry. Finally, meeting the requirements of increased regulation has forced shipbuilders to acquire additional tools and develop new processes, which have added to their overhead costs.

CHAPTER FIVE

Options for the Navy to Reduce Ship Costs

What can the Navy do to reduce its ship costs? While this study did not conduct an exhaustive search of ways that these costs could be reduced, our interviews with shipbuilders and other knowledgeable sources elicited a dozen preliminary ideas related to the issues we found:

- Increase investments in shipbuilding infrastructure aimed at improving producibility
- Increase shipbuilding procurement stability
- Fund shipbuilding technology and efficiency improvements
- Improve management stability
- Change GFE-program management controls
- Employ batch production scheduling
- Consolidate the industrial base
- Encourage international competition and participation
- Build ships as a vehicle
- Change the design life of ships
- Buy a mix of mission-focused and multi-role ships
- Build commercial-like ships.

Some of these ideas are highly speculative and, given the current fiscal and legislative environment, have dubious prospects for implementation. Nonetheless, we present all of them for completeness of discussion, addressing each and pointing out both positive and negative aspects.

Increase Investments in Shipbuilding Infrastructure Aimed at Producibility

Although instability in U.S. naval shipbuilding planning discourages U.S. shipbuilders from making investments to improve their efficiency, shipbuilders elsewhere provide many examples of how investments can improve efficiency. Many European and Asian shipbuilders have developed new and improved techniques for building commercial ships, including cruisers, tankers, bulk carriers, and high-speed catamaran ferries (Hess et al., 2001). While not all these ideas can be directly applied to military ships, some U.S. naval shipbuilders have already translated certain techniques, such as modular berthing assemblies, into naval ships. Increased capital investments could also help to reduce the need for workers in this labor-intensive industry, thereby addressing some of the workforce problems naval shipbuilders face.

Investments could come directly from the Navy seeking to reduce its future ship costs, from the U.S. government as part of its economic or defense industrial base policies, from state or local governments seeking to stabilize their local economies, or from the private sector if the return on investment were appropriate. Because the shipbuilding industry involves heavy manufacturing, the required investments are not likely to be small. Furthermore, many may view government investment in shipbuilding as inappropriate. We do not seek to address the question of whether government investment would be appropriate but note that industry representatives apparently believe the return on investment to be insufficient for private financiers given the uncertainty and paucity of naval ship orders.[1]

---

[1] Much of this is discussed in the *Global Shipbuilding Industrial Base Benchmarking Study—Part I: Major Shipyards* (2005), which was completed under the auspices of the Office of the Deputy Under Secretary of Defense (Industrial Policy). The study finds that U.S. industry best practice ratings increased between 2000 and 2005 by roughly 15 percent. The study recommends that DoD establish a shipbuilding industrial base investment fund to help U.S. shipyards further increase their productivity.

Increase Shipbuilding Procurement Stability

As noted previously, the Navy varies the number of ships it buys each year. The current, FY 2006 Navy Shipbuilding budget requests only four ships—a stark contrast to the roughly 20 ships per year it procured in the 1980s (not including landing craft). Its current 30-year shipbuilding projection includes plans for achieving a fleet ranging in size from 260 to 376 ships, a broad range corresponding to the uncertainty in the number of ships that will ultimately be purchased. As also noted, this uncertainty causes shipbuilders to be hesitant about investments to increase productivity through cost-saving devices or methods.

Multiyear buys, or "advance appropriation," could introduce a greater semblance of stability to naval shipbuilding (Blickstein and Smith, 2002). The Navy could also improve stability by resisting changes to its shipbuilding plan with each budget cycle. Below, we review the prospects for multiyear procurement, incremental procurement, and advance appropriation.

Multiyear procurement has been used primarily in aircraft procurement (e.g., F/A-18E/F) but has also been used for DDG-51 and SSN-688 ships. This procurement strategy permits the Navy and shipbuilders to establish contractual agreements for future ships over several years. Under multiyear procurement, Congress authorizes all the procurement quantities and funding necessary in the first year of a multiyear procurement but only appropriates the funds necessary for the first year of the procurement. It then appropriates funding in each subsequent year as part of the annual DoD appropriation bill. Because multiyear procurements also establish penalties against the Navy for not procuring the specified number of ships, and because Congress rarely backs away from such an agreement once it is a matter of law, such agreements give shipbuilders greater confidence in making investments and also allow them to increase their purchasing leverage with suppliers.

Incremental procurement, in which portions of ships are appropriated by Congress on a year-by-year basis, has been used to buy LHD-8 ships. Typically, individual ships must be fully funded (appropriated) in the year of authorization, even though these funds will be paid to the contractor over several years as the ship is constructed. In incremental procurement, funds are appropriated annually for each year of the ship's construction period, thus spreading out the procurement dollars over many years, providing some money up front and the knowledge that future Congresses will likely add funding to avoid leaving a partially completed ship in a shipyard. Some advocate using this technique for large and expensive ships, such as aircraft carriers. DoD has occasionally supported this concept to reduce the large amount of money appropriated for one ship in a given fiscal year and to avoid the reduction in the procurement of other ships. While the Navy and contractors tend to support incremental procurement, Congress has traditionally opposed this type of procurement because it saddles future Congresses with a procurement plan that may not be desirable in the future.

Advance appropriation is similar to incremental procurement but requires that Congress appropriate funds for ensuing fiscal years in the first year of the program. The difference between advance appropriation and incremental appropriation is that Congress acts with knowledge and forethought in the case of advance appropriation and commits future Congresses to the funding pattern. In the case of incremental funding, monies are added on a year-by-year basis, and the shipyards have no assurance that the project, once started, will be continued. Advance appropriation has many of the positive features of incremental funding but requires a specific appropriation by Congress for each of the ensuing funding years. It also has the disadvantage of obligating future Congresses. Although the Office of Management and Budget permits advance appropriation in its regulations, it discourages its use.

Each of the above techniques is advantageous to the shipbuilding industry because it provides some advance notification of future procurement by the Navy—and thereby permits shipbuilder planning and investment—that cannot be easily changed, even in the face of increasing budget pressures.

Fund Shipbuilding Technology and Efficiency Improvements

The Navy could fund shipbuilding technology improvements through its research and development arm, the Office of Naval Research. Such a program would foster research to reduce costs through efficiency gains realized with new and better technologies. Such programs would, in effect, constitute an industrial policy in which the federal government would be singling out military shipbuilding as an industry it desires to protect by increasing its productivity. While there have been programs, such as MARITECH and the National Shipbuilding Research Program, in the past to fund research to increase shipyard productivity, questions of how much these programs have reduced the cost of ships to the Navy still remain.

Improve Management Stability

One possible way the Navy could control customer-driven factors in ship cost escalation is through improved management stability. For example, the Navy could curtail its current practice of rotating officers through jobs every three to four years. Because a ship may take up to seven years to build, a program could have as many as three program managers, each with differing management methods and philosophies.

Management initiatives could also help reduce the number of change orders during the production. While change orders may be driven by changes in the military threat environment, experience with earlier ships of a given class, or changes in environmental regulations or quality of life considerations, change management is still in the hands of Navy leadership. Changes to military vessels average about 8 percent of total procurement cost (Arena et al., 2005). During the early years of the Reagan administration, the Secretary of the Navy insisted that he approve every change to a ship or aircraft contract—taking away, in effect, such decisions from acquisition officials and transferring it to his office. While drastic, such a measure can control change orders.

Change GFE Program Management Controls

Some argue that the Navy ship program manager has little control over the equipment or systems that are purchased and installed on a ship, but rather must deal with GFE managers for systems such as tactical radios, radar, sonar, and antennae. This makes the ship program manager subject to the schedule and cost constraints of those programs. In the opinion of some shipbuilders, giving the ship program manager more control over the cost and schedules of GFE products would also help give the shipyards better control of their schedules by having a single point of contact for both the ship construction and the GFE delivery, thereby improving cost control. Others contend that this would suboptimize the GFE programs and ultimately lead to increased costs for them. Some also contend that the GFE management should be transitioned to the shipbuilders. Schedules for different portions of shipbuilding do not always mesh with the GFE, leading to increased costs due to involving more managers in the process and schedule delays.

If shipbuilders were to write GFE controls, schedules might mesh better, although this could entail more costs for the government, requiring it to pay a fee in addition to the price charged by the GFE vendor and some management charge by the shipbuilder. Some initial attempts to use such a process have met with mixed success, including the integrated area networks on CVN-77 and LPD-17. Coordinating common systems across multiple programs would likely need to remain a government function.

Employ Batch Production Scheduling

Batch production scheduling, used in other government procurement programs, could be adapted to shipbuilding, with production of a given type of ship concentrated in a shorter period of time. For example, the Navy could buy only submarines for a specified period of time, then shift to surface combatants, then build amphibious ships, then construct submarines again. This would increase the rate of production for a given type of ship during a specified time and perhaps realize learning benefits, economies of scale, and other benefits. If this were coupled with incremental funding, advance procurement, or other budget-stabilizing techniques, shipyards may also be further encouraged to make investments that would ultimately reduce unit costs. The Air Force has used this technique to concentrate production for transport aircraft, fighters, and tankers. One difficulty of adapting this approach to naval ships lies in the specialization of shipyards in specific types of vessels, a specialization not matched by aircraft manufacturers (Birkler, Bower, et al., 2003). If, for example, the Navy were to concentrate its shipbuilding in this way, individual shipyards would likely have no work for an extended period of time and would be unable to retain the skills of their design and production workforces. This, in turn, would mean that the government would likely incur significant costs when seeking once again to construct vessels.

Consolidate the Industrial Base

A recurring issue since the end of the Cold War has been the rationalization of the defense industrial base to fit current budget capacity. Considerable consolidation has occurred within shipbuilding, with the loss of a shipyard in Baltimore; Northrop Grumman's acquisition of yards in Louisiana, Virginia, and Mississippi; and General Dynamics' acquisition of yards on the West Coast. The United States now has only one contractor for aircraft carriers and two each for surface combatants, submarines, and amphibious ships. The Navy could encourage further consolidation, allowing the surviving yards to construct more ships but eliminating any semblance of competition. Although competition might help reduce prices, there is also little evidence, as noted previously, that current "allocation" processes gains the benefits of competition.

Encourage International Competition and Participation

For security reasons, U.S. Navy ship production is limited to U.S. producers. While this makes sense from a defense perspective, it limits competition and innovation that might be realized from procuring ships in a global market. Allowing the Navy to buy from foreign companies could help increase competition and reduce costs. Some of our NATO allies currently build both military and commercial ships. Competition or participation by foreign suppliers might have a positive effect on the U.S. shipbuilding industry. The globalization of the defense industry is, in general, a process that is evolving. For example, BAE Systems, a UK company, is participating in the development of the Joint Strike Fighter. Foreign companies are also increasingly buying U.S. companies. The purchase of United Defense by BAE Systems is a general example of how foreign firms are helping to bring or maintain the benefits of increased competition and reduced costs in a U.S. industry. Still, involving foreign firms in U.S. naval shipbuilding could raise issues of access to and control of ship technology.

Build Ships as a Vehicle

Building the ship as a vehicle could help to reduce costs by separating the mission systems from the HM&E (hull, mechanical and electrical). A ship could be purchased as a "bus" that would permit others to install GFE. Under such a system, the shipbuilder may be responsible only for the hull and mechanical and electrical equipment and then turn the ship over to the government to install GFE. The government would have to assume the roles of test, evaluation, and integration, which are currently performed by the contractor. The government would then have GFE contractors install equipment at the dock. Such a system would relieve the shipbuilder from concerns over delivery and installation of equipment provided by others, transferring it to the government. Another disadvantage of such a system is that it would likely extend the time necessary to provide a "ready" ship to the Navy. Another concept, used by the Navy in the DD-963 program, would be to purchase "large" and relatively "empty" ships. Systems and equipment were added to DD-963 ships as threats or capabilities arose. This was, in a sense, an early version of "spiral" development for the Navy. The FFG-7 class followed a similar evolutionary approach. Such an approach, however, would likely increase total outfitting hours as it is much more difficult to outfit a ship that is fully assembled.

Change the Design Life of Ships

Ships could be designed to last a shorter, or longer, period of time. Currently, the Navy expects ships to have at least a 30-year period of life. Given the enormous changes in technology during a 20-year span, it might make sense to build ships for a 20-year period of life. One could couple this concept with a "no-change" policy, pushing changes to the next class of ships. RAND has researched this possibility for aircraft carriers, suggesting that the Navy might desire to build new ships rather than refuel older carriers (Schank, Smith, et al., 2005). Alternatively, the Navy might seek to have ships with a period of life exceeding 50 years. This could entail different, more modular methods of ship construction, with portions of ships built to last a long time. Further analysis would be required to determine whether either of these concepts is practical.

Buy a Mix of Mission-Focused and Multi-Role Ships

The Navy could buy a mix of ships—some specialized to a particular mission and others that could serve multiple roles within the fleet. This approach has been sometimes called a "high-low" mix. The cost advantage of mission-focused ships over the multi-mission ones is that the mission-focused ships require fewer systems and are easier to integrate. Such hulls would likely be smaller as well. In fact, it appears that the Navy is currently pursuing such a procurement strategy. Current procurement plans call for DD(X) on the multi-mission side and the LCS on the mission-focused side. The Navy also used this strategy in the late 1970s when it began production of FFG-7 ships. Originally designed as the mission-focused ship, the FFG-7 (as noted earlier) was upgraded and given multiple missions over time, evolving from what had been a relatively simple antisubmarine warfare ship.

Build Commercial-Like Ships

Some navies build ships to more commercial architectural and equipment standards than the U.S. Navy. These ships have different survivability goals, focusing on saving the crew rather than ensuring a ship could continue to fight after damage. For example, the UK Royal Navy built HMS *Ocean* (an amphibious vessel) using a commercial approach and standards. The Royal Dutch Navy builds surface combatants that are more similar to commercial ships in their survivability and build standards. Such ships are less expensive and easier to build than military ones. Such ships are also more likely to be lost after an attack. The USS *Cole*, for example, likely would have sunk in the Gulf of Aden if it had been built to a lower standard of survivability.

Summary

As we have discussed in this chapter, there are myriad ways to reduce ship costs. Some of these are not useful or executable. For example, competing naval ships in the international market would probably not be permitted by DoD or Congress. Others may have only a marginal value in reducing costs. Our earlier analyses show that increases in labor, material, and equipment costs are roughly comparable to other inflation indexes, meaning that pressure to reduce costs to shipbuilders, while a laudable goal, will likely not be enough to solve cost escalation problems for naval ships. Our analyses have also shown that neither increasing nor decreasing procurement rates is likely to have a great effect (i.e., one exceeding 10 percent) on the cost of ships.

The factor that appears to be the greatest contributor to the increasing cost of ships is changing requirements. While the nation and the Navy understandably desire technology that is continuously ahead of actual and potential competitors, this comes at a cost. We do not evaluate whether the cost is too high or low; we note only that it exists.

The Navy appears to be moving toward a high-low procurement strategy, with construction of the LCS on the low side and the DD(X) on the high side. For its amphibious ships, it seeks to reduce costs of both the new-design LHA (amphibious assault ship) and the commercial variant MPF(F) cargo-like ship. Given that the pressures on shipbuilding funds will continue in the foreseeable future, the Navy will continue to seek ways to reduce the costs of its ships.

CHAPTER SIX
Conclusion

As Admiral Vernon Clark, former CNO of the Navy, noted in his testimony to Congress, the cost escalation for naval ships has exceeded most measures of inflation. Our analysis found that naval ship costs have escalated at a rate between 7 and 11 percent since the 1950s. Common measures of inflation over that time period ranged from 4 to 5 percent; thus, the cost increases for naval ships have substantially outpaced inflation. Such a rapid cost escalation has significant implications for the Navy and the composition of its fleet given that shipbuilding budgets are not likely to increase for the foreseeable future.

To understand the causes of this growth, we categorized the cost escalation into economy-driven and customer-driven factors. The economic factors include influences of price changes largely outside the government's control. These factors include direct labor rates, other labor costs (e.g., benefits), material and equipment costs, and productivity. Labor costs grew faster than inflation (driven by increases in indirect costs), whereas material and equipment costs grew less than inflation. Overall, we found that increases for these economic factors accounted for about half the overall escalation and were similar to common indexes for inflation (e.g., CPI, GDP, and DoD deflator).

The customer-driven factors are those that the customer (both the federal government and Navy) directly influences. One example of a Navy-driven factor is the capabilities and features of the ship that are reflected in the complexity (e.g., size, speed, power generation, stealth, survivability, habitability, and the number of mission and armament systems). There are other customer-driven factors that are related to how ships are built, such as worker safety, production rates, and procurement strategy that may be outside the Navy's control but are driven by policy from the federal government (e.g., OSHA worker regulations). We found that the customer-driven factors accounted for the remaining half of the ship cost escalation. Complexity and features accounted for most of the increase, while production rates and other influences contributed a lesser amount. It should be noted, however, that some of the cost increases due to decreases in production rates were confounded with changes in economic factors (namely, indirect rates). We were unable to separate the influences of rate from these factors.

We explored a number of options that could be considered to reduce or contain the escalation in ship costs. These options ranged from those designed to contain requirements growth to those designed to improve production productivity. All these approaches require some level of compromise or investment. No simple solutions exist. A combination of these efforts, including those designed to contain ship requirements (e.g., the LCS-like approach of separating the ship from the mission systems or building mission-focused ships) and some investment in improving shipbuilding efficiency, may be most appropriate.

Improving ship capability and effectiveness is not, of course, a bad thing. While the costs for ships have increased, so has the effectiveness of our fleet. Our technological superiority can deter and overwhelmingly defeat our adversaries. When shipbuilding budgets were larger and increasing, such cost increases were not a problem. Now, with tighter and potentially diminishing shipbuilding budgets, we need to make harder choices between the capabilities and numbers of ships.

Ship Classes Included in the Multivariate Regression Analysis

The ship classes listed below are those used in the multivariate regression analysis of ship characteristics described in Chapter Two. Note that two of the 37 observations provided by NAVSEA were not included in the analysis. The CVN-68 was excluded because it was the only aircraft carrier observation in the sample. The observation did not fit the regression well, indicating that an offset term (similar to the ones for submarines and auxiliaries) is possibly needed. However, with one observation it is not possible to determine the correct offset term. The AGOS-23 was also dropped because of its very different hull form (a swath vessel).

• AD-41	• AO-177	• FFG-7
• AGOR-21	• AO-187	• LHD-1
• AGOR-23	• AOE-6	• LSD-41
• AGOS-1	• ARC-7	• LSD-49
• AGOS-19	• ARS-50	• MCM-1
• AGS-39	• AS-39	• MHC-51
• AGS-45	• ATF-166	• MSH-1
• AGS-51	• CG-47	• PHM-1
• AGS-60	• CGN-38	• SSBN-726
• AKR-295	• DD-963	• SSN-21
• AKR-300	• DDG-51	• SSN-688
• AKR-310	• DDG-993	

Multivariate Regression for Ship Cost

Table B.1 (on the following page) lists the multivariate regression analysis for ship cost as a function of various complexity characteristics. Note that there are more than 35 observations (the number of individual ship classes), since each shipbuilder, class, and flight was treated as a separate observation; therefore, a ship class might have multiple observations. (For example, the DDG-51 that is built by both Northrop Grumman and General Dynamics has evolved through three flights. Thus, there are six observations for the DDG class.)

Table B.1
Multivariate Regression Output for Ship Characteristics

Number of obs = 41
F(4, 36) = 96.29
Prob > F = 0.0000
R-squared = 0.9145
Adj R-squared = 0.9050
Root MSE = 0.33965

Source	SS	df	MS
Model	44.4351384	4	11.1087846
Residual	4.15313426	36	0.115364841
Total	48.5882726	40	1.21470682

lnc9	Coef.	Std. Err.	t	P > |t|	[95% Conf. Interval]	
lnlsw	0.9469314	0.0686195	13.800	0.000	0.8077645	1.086098
lnpowden	0.9434869	0.1826812	5.165	0.000	0.5729923	1.313981
aux	−1.288984	0.1259019	−10.238	0.000	−1.544325	1.033643
sub	0.2896558	0.1528438	1.895	0.066	−0.0203259	0.5996375
_cons	11.49907	1.188007	9.679	0.000	9.089678	13.90846

Where

- lnc9 is the natural log of cost for the hypothetical ninth unit (fit by a cost improvement curve) in FY 2005 k$
- lnlsw is the natural log of the LSW in tons
- lnpowden is the natural log of the power density in kW per ton (LSW)
- aux is a binary term for an auxiliary vessel (1 = true, 0=false)
- sub is a binary term for a submarine (1 = true, 0 = false).

RAND Questions to Each Firm

We posed the following questions to each shipbuilder:

1. In your view, what are the primary sources of ship cost escalation for the past several decades?
2. How has the complexity of ships evolved? What metrics do you think capture best the evolution in the complexity of ships (e.g., power generation, weapons on board, number of days at sea, number of support equipment, area of regard)?
3. Are there any changes to contractual, regulatory, and statutory requirements that you believe may have added to the acquisition costs of ships for the past six decades (e.g., contract military specification requirements, OSHA requirements, environmental requirements)? If so, how can we quantitatively or qualitatively capture or reflect their effects on ship costs?
4. Overhead costs have grown in the past few decades. Do you have any data that illustrate the increase in overhead costs? Can you provide the main contributors to the overhead cost growth (e.g., retirement, health care, other benefits)?
5. Please tell us about the changes in industrial base that may have affected ship costs (e.g., diminishing sources of materials and equipment, lack of competition at the sub-vendor level, flow-down of government requirements).
6. Are there any government reporting requirements that have changed in the past six decades (e.g., earned value, cost data

reporting, CDRL [Contract Data Requirements List] requirements)? Do you believe that any of these requirements have
added to the ship costs?

7.   Are there any disincentives in how the government procures
ships that may lead to cost growth? Are there any initiatives
that the government can encourage to reduce the cost of future
ships (e.g., multiyear acquisition, lean production, contractual
incentives for cost reduction)?

Cost Escalation Over the Past 15 Years

During an initial briefing of this analysis, a question arose as to whether the results would look different if a more recent time frame was chosen. In other words, rather than looking at trends over the past 40 to 50 years, one should look at the trends over the past 10 to 15 years. The argument asserted here is that the issues confronting the industry today are not the same as those a few decades ago. While that may be true, this study's objective was to examine long-term trends. Nonetheless, in this appendix we will show the trends for similar metrics over the past 15 years (1990–2004). Overall, our general conclusions do not change.

Historical Comparisons

While it is relatively easy to show trends with the escalation indexes for labor, material, etc., over a short time frame, it is much harder to examine the trends for naval shipbuilding. A 15-year window is relatively short by typical ship program life cycles (and for most weapon systems). For example, there has not been a new surface combatant class produced during that time frame. For other ship types, such as attack submarines, there have been new classes designed and produced. Exploring a rate of change will be strongly affected by when a new class (if any) was introduced in the 15-year window. Therefore, any values for the rate of escalation should be looked as indicative and not exact. Furthermore, we have added estimated costs for some of the current classes in production (e.g., LPD-17, SSN-774, CVN-21) based on

Shipbuilding and Conversion, Navy (SCN) budgets (rather than final values as used in the main body of this report). This addition of data was done to fill in the past few years for which we lacked recent observations. Table D.1 summarizes the cost escalation rate for surface combatants, submarines, carriers, and amphibious ships over the 1990–2004 time frame. As before, these rates are *unadjusted* for inflation.

The surface combatant growth rates look low in comparison to the other ship types because there was not a new class introduced in the time frame. We did not have cost information to add a DD(X) estimate to the sample. The attack submarine rate is lower, since there was a scaling back (smaller and less complex) between the *Seawolf* and *Virginia* classes in a deliberate attempt to make a more affordable submarine. The trend for carriers is about the same, showing a steady evolution of the class. Despite these differences, the range is very similar (albeit broader) to the analysis in the main text: 3 to 12 percent annual growth rate. The annual growth rate for fighter aircraft in the 1990–2004 time frame was 8 percent, which again was similar to the ships' growth rate.

In Table D.2, we summarize the cost escalation for the CPI (and some of its components), GDP deflator, and the DoD procurement deflator from 1990 to 2004. Note that in the 15-year time frame, the DoD procurement deflator is lower than any of the indexes—even the GDP deflator. The difference suggests that forecasts of inflation for ongoing and future programs might be too low. Figure D.1 shows the relative trend among the three indexes.

Table D.1
Battle Force Cost Escalation Rates, 1990–2004

Ship Type	Annual Growth Rate
Amphibious ships	12.0
Surface combatants	3.4
Attack submarines	4.8
Nuclear aircraft carriers	7.1

Table D.2
Annual Growth Rate for Comparison Indexes,
1990–2004

Index	Annual Growth Rate (%)
DoD procurement deflator	1.6
GDP deflator	2.0
Private transportation (new and used cars, repairs, fuel, etc.)	2.0
CPI	2.7
Medical care	4.7
College tuition	6.8

Figure D.1
Comparison of DoD and GDP Deflators with the CPI, 1990–2004

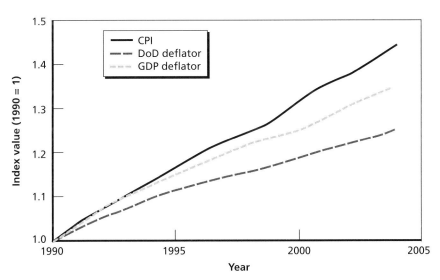

RAND *MG484-D.1*

Points of Comparison

Selecting comparison points over the broader time frame was easy, since there was typically more than one similar ship class produced (e.g., a DDG, SSN, LPD, or CVN). However, over the shorter 15-year period such a choice becomes difficult, if not impossible. One exception was the amphibious ships, for which three very different classes were produced over the time period (the LHD-1 and LSD-41 classes were produced in the early to mid-1990s, and the LPD-17 class started in the late 1990s). Only one surface combatant class was built, the DDG-51. Similarly, only one carrier class was built, the CVN-68 class. Therefore, looking at specific differences in ship characteristics and quantifying escalation will be difficult. To again assess and validate the magnitude of the various escalation factors, we will need to make comparisons between specific classes and not just general time frames. The comparison points we have chosen for this 15-year time frame analysis are:

- Attack submarine: SSN-768 (FY 1988) to SSN-777 (FY 2002)[1]
- Surface combatant: DDG-51 (FY 1989) to DDG-51 (FY 2002)
- Aircraft carrier: CVN-75 (FY 1988) to CVN-76 (FY 1995).

We have not included an amphibious example because we felt that a specific comparison did not prove meaningful—that is, the types of ships were too dissimilar.

Economy-Driven Factors

As we did previously, we split the economy-driven factors contributing to cost escalation into labor, equipment, and material. Figure D.2 shows the average annual growth rate for shipyard labor (burdened and unburdened) for the Big Six shipyards over the 1990–2004 time frame.

---

[1] The SSN-768 is a 688i (*Los Angeles*–class) submarine, and the SSN-777 is a 774 (*Virginia*-class) submarine.

Figure D.2
Shipbuilding Labor Rate Escalation, 1990–2004

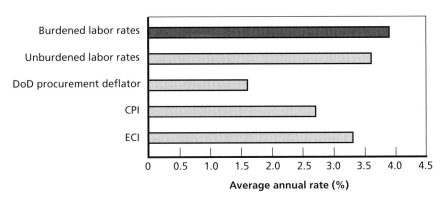

RAND *MG484-D.2*

There are some important differences to note. One such difference is that both the burdened and unburdened rates grew faster than the CPI. Another difference is that the direct rate grew slightly faster than the ECI, indicating that shipbuilding direct labor costs have escalated slightly more than other manufacturing industries have. As was discussed in Chapter Four, the greater direct growth rate could be due to the fact that the shipbuilders have to pay a premium to retain skilled workers and that average wage rate per worker is increasing because the workforce comprises very experienced workers (who draw higher wages). Again, the DoD procurement deflator is about half that of the other indexes. As others have noted, the procurement deflator does not appear to accurately mimic cost escalation for shipbuilding (Deegan, 2004).

In Figure D.3, we show the same six material and equipment indexes described in Chapter Three:

1. Electrical machinery and equipment (BLS series ID: WPU117)
2. Electronic components and accessories (BLS series ID: WPU1178)
3. Switchgear, switchboard, relays, etc., equipment (BLS series ID: WPU1175)

4.   Mechanical power transmission equipment (BLS series ID: WPU1145)
5.   General purpose machinery and equipment (BLS series ID: WPU114)
6.   Steel mill products (BLS series ID: WPU1017).

For comparison purposes, we also show the indexes for the annual growth for the GFE electronic equipment identified in the P-5 break-out from the SCN budgets for DDG-51 (1997–2004) and SSN-774 (1998–2004).

Interestingly, both of the indexes for electronic components (numbers 1 and 2) show a decline (deflation) over the 15-year period. The other components show gains significantly greater than that of the DoD deflator but less than the CPI. The growth rates for both GFE electronics examples were less than the DoD deflator. The average calculated rates for material and equipment escalation between 1990 and 2004 are shown in Table D.3. Notice that the calculated rate for equip-

Figure D.3
Material and Equipment Cost Escalation, 1990–2004

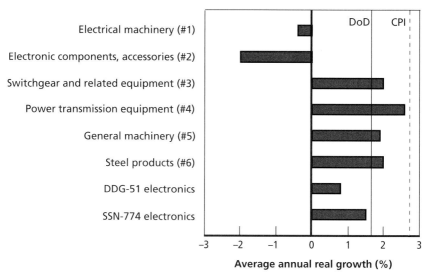

Table D.3
Material and Equipment Annual Escalation Rates,
1990–2004

Commodity	Annual Escalation Rate (%)
Material	2.0
Equipment	1.6

ment escalation is nearly identical to that for the SSN-774 GFE electronics escalation.

Customer Factors

Changes due to the characteristic complexity factors (LSW and power density) over the brief time frame are difficult to quantify. We do not have power density information at the hull level—only the class level. For the *Nimitz* class, we assumed that the changes were negligible. For the attack submarines (SSN-688i to SSN-774), there are changes in displacement and power density, although these are much smaller compared with the SSN-637 as a baseline. For the DDG-51s, the difference in LSW over the class was about 4 percent. Again, we do not have data for the changes in power density within a class.

As before, we used the 1998 Electric Boat study of the changes in construction cost for recent classes of attack submarines. The rates were calculated as before, but the baseline is now the SNN-688i class and not the SSN-637. The rates for cost escalation due to standards, regulations, and requirements are listed in Table D.4.

Table D.4
Cost Escalation Due to Standards, Regulations,
and Requirements

Vessel Type	Annual Escalation Rate (%)
Submarines	2.0
All other ships	1.7

Total Contribution of Factors

As before, we will summarize the contributions of the various factors into three broad categories and compare the total with the actual escalation rate for the specific ship comparisons. The economy-driven category comprises the changes due to labor cost, material, and equipment price changes. The customer-driven category comprises changes due to characteristic complexity, standards, regulations, requirements, and procurement rates. A third category corrects for specific hulls being at different points on the cost improvement curve. Table D.5 displays these results.

Two of the three totals (surface combatant and aircraft carrier) are fairly close, given the error of the method, to their actual values. The predicted submarine value is much higher than the actual one. One possibility is that we have overcorrected for the change to cost improvement for submarines.

Summary

In this appendix, we have explored the sensitivity of time frame to our analysis. Specifically, we chose a shorter and more recent time frame—the past 15 years. As before, we were able to forecast the cost escalation of ships over this period and found them to be close to the actual rates. The annual rates for the component indexes were lower for the 10- to 15-year

Table D.5
Contributions to Annual Escalation Rate by Customer-Driven Factors

Ship Type	Economy-Driven Factors (%)	Customer-Driven Factors (%)	Cost Improvement Correction (%)	Total (%)	Actual (%)
Surface combatant	2.4	2.3	−0.9	3.8	3.5
Attack submarine	2.6	3.6	1.8	8.0	6.5
Aircraft carrier	2.3	1.8	0.0	4.1	4.5

period than for the 40- to 50-year period. In addition, the DoD procurement deflator was much lower than any of these rates over a similar time frame. However, economy-driven and customer-driven factors were roughly equal in magnitude.

Passenger Ship Price Escalation

An extension of the escalation analysis, which was suggested by Larrie Ferreiro, professor of systems engineering at Defense Acquisition University, was to compare the cost escalation for commercial vessels with naval ones. In this appendix, we will analyze the price escalation for passenger (cruise) ships ordered between 1997 and 2005 and compare their results with those for the Navy ships.

Data

Dr. Ferreiro provided to RAND a data set of approximately 230 passenger ships ordered between 1994 and 2005. These data were based on information published in *SeaTrade Cruise Review* between 1997 and 2005. The information contained the following elements:

- Price of the vessel
- Owner
- Builder
- Order date
- Delivery date
- Name of the vessel
- Gross registered tons (GRT)
- Number of lower berths.

Table E.1 lists the summary statistics for these data.

Table E.1
Statistical Summary of Passenger Ship Data

Element	Number of Observations	Mean	Minimum	Maximum
Year of order	227	1999.2	1994	2005
Number of lower berths	225	2,040	96	3,840
GRT	227	81,600	3,000	160,000
Price (millions $)	227	375.2	23	828

One important caveat to these data is that the values are those reported in the open literature. So, the accuracy and validity of the information is questionable, particularly with respect to price. We have made no attempt to validate them. Furthermore, all the prices were reported in U.S. dollars and based on a conversion practice that was not made known to the reader. How the numbers were converted to dollars could influence prices, particularly where the dollar has been weak against European currencies recently. Therefore, we cannot exclude some effect due to exchange rate trends.

Analysis

The annual escalation rate for this data sample was 8.9 percent, a rate similar to naval ships. Such a rate is somewhat unexpected given that the product is a "commercial" one. The typical perception is that commercial products have not experienced as much cost escalation as military products have. However, these increases in price can be explained mostly in the changes in the size of the ships being produced. Over the time frame of the data, the sizes of passenger ships have grown substantially. Figure E.1 shows this size trend.

Figure E.1
Passenger Ship Size vs. Year of Order

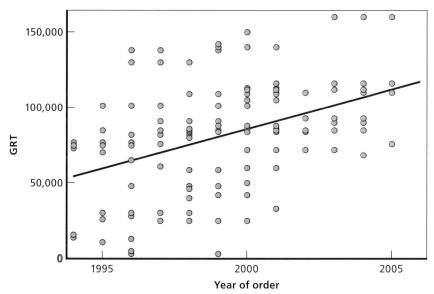

To adjust for this trend of increasing size, we determined a relationship between price (in a fixed 2005 dollar basis) and size by least squares regression. This regression relationship is as follows:

$$\ln(\text{price05}) = 0.609\ln(\text{grt}) - 0.837, \qquad (\text{E.1})$$

where ln(price05) is the natural logarithm of price in millions of FY 2005 dollars,[1] and ln(grt) is the natural logarithm of the GRT. Figure E.2 shows a plot of the relationship.

---

[1] We used a European average GDP deflator to convert the prices to the constant FY 2005 basis. Realistically, the costs should be adjusted using country-specific rates. However, the level of this analysis did not warrant such specificity.

Figure E.2
Regression Relationship Between Price and Gross Registered Tonnage for Passenger Vessels

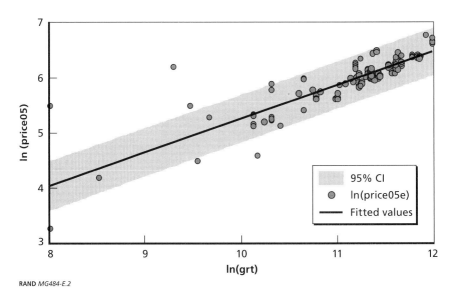

RAND *MG484-E.2*

In Figure E.2, the shaded region of the plot is the 95 percent confidence interval for the forecasted values, the line represents the fitted values, and the points are the actual observations. Table E.2 shows the full regression diagnostics.

Table E.2
Regression Diagnostics for ln(price05) vs. ln(grt)

Number of obs = 227
F(1, 225) = 688.21
Prob > F = 0.0000
R-squared = 0.7536
Adj R-squared = 0.7525
Root MSE = 0.21659

Source	SS	df	MS
Model	32.2858264	1	32.2858264
Residual	10.5553643	225	0.04691273
Total	42.8411907	226	0.189562791

ln(price05)	Coef.	Std. Err.	t	P > |t|	[95% Conf. Interval]	
ln(grt)	0.6088125	0.0232072	26.23	0.000	0.5630813	0.6545438
_cons	−0.8365517	0.2597134	−3.22	0.001	−1.348333	−0.32477

Note that we also explored including a term in the regression related to the number of lower berths on the ship. As might be expected, the number of lower berths and GRTs were highly correlated (>0.8 correlation) and was, therefore, omitted from the final regression. A term for the number of lower berth per ton was not statistically significant (ignoring two leveraging observations).

The statistics are not good enough to consider this equation a formal cost estimating relationship, because of high residual error. Its purpose is to scale the cost for differences in the size of passenger vessels, not forecast price. For example, the relationship does not work as well for ships with GRT below about 20,000 tons because of higher residuals for those observations. Also, there are many more ship factors one would wish to consider in an estimating relationship (power density, fit and finish quality, etc.). The important result from this analysis is that the scaling coefficient is less than one, indicating some economies of scale for larger ships. Recall that the scaling coefficient for the naval ships was nearly equal to one.[2] If we control for size using Equation E.1, the annual escalation rate for the passenger ships becomes 2.9 percent. This value is more in line with the European GDP deflator annual growth rate of 2.2 percent over the same time frame.

It is interesting that there remains a 0.7 percent difference for which we cannot account. Again, "hidden factors" might be driving this increase. Notably, passenger ships have also undergone improvements in standards for crew and passenger safety, environmental protection, complexity of entertainment systems, and so forth. This escalation is about a third of that observed for the U.S. naval ships. However, some of the true increase might be masked by productivity gains in manufacturing passenger vessels. Without specific data on these improvements and productivity, the comparison is only speculative.

Summary

We have examined the escalation rates for passenger ships ordered between 1994 and 2005 and found that their reported prices have escalated by a similar rate to the U.S. naval ships—about 9 percent per year. This escalation rate is primarily a result of the increase in the size of the ordered ships (customer-driven). The remaining increase appears to be slightly higher than normal price escalation for the manufacturing countries. These results further support our conclusion that escalation beyond inflation is driven mainly by the demand for greater capability and performance.

---

[2] Admittedly, the two formulations use very different measurements of size. The passenger ships' measure of GRT is a volumetric measurement, whereas the naval ships' measure is weight based (LSW).

Bibliography

Alexander, Arthur J., and Bridger M. Mitchell, "Measuring Technological Change of Heterogeneous Products," *Technological Forecasting and Social Change*, Vol. 27, Nos. 2–3, May 1985, pp. 161–195.

Arena, Mark V., John Birkler, John F. Schank, Jessie Riposo, and Clifford A. Grammich, *Monitoring the Progress of Shipbuilding Programmes*, Santa Monica, Calif.: RAND Corporation, MG-235-MOD, 2005.

Augustine, N. R., *Augustine's Laws*, New York: Viking Penguin, 1986.

Berenson, Mark L., and David M. Levine, *Basic Business Statistics: Concepts and Applications*, 6th edition, Englewood Cliffs, N.J.: Prentice Hall, 1996, pp. 824–825.

Birkler, John, John C. Graser, Mark V. Arena, Cynthia R. Cook, Gordon Lee, Mark Lorell, Giles Smith, Fred Timson, Obaid Younossi, and Jon G. Grossman, *Assessing Competitive Strategies for the Joint Strike Fighter: Opportunities and Options*, Santa Monica, Calif.: RAND Corporation, MR-1362-OSD/JSF, 2001.

Birkler, John, Anthony G. Bower, Jeffrey A. Drezner, Gordon Lee, Mark Lorell, Giles Smith, Fred Timson, William P.G. Trimble, and Obaid Younossi, *Competition and Innovation in the U.S. Fixed-Wing Military Aircraft Industry*, Santa Monica, Calif.: RAND Corporation, MR-1656-OSD, 2003.

Birkler, John, Denis Rushworth, James Chiesa, Hans Pung, Mark V. Arena, and John F. Schank, *Differences Between Military and Commercial Shipbuilding: Implications for the Untied Kingdom's Ministry of Defence*, MG-236-MOD, Santa Monica, Calif.: RAND Corporation, 2005.

Blickstein, Irv, and Giles Smith, *A Preliminary Analysis of Advance Appropriations as a Budgeting Method for Navy Ship Procurements*, Santa Monica, Calif.: RAND Corporation, MR-1527-NAVY, 2002.

Bruno, Michael, "Navy Acquisition Secretary to Recommend One DD(X) Shipyard," *Aerospace Daily & Defense Report*, March 11, 2005a.

———, "CBO: Navy's Fleet Plan Requires $15B a Year in Shipbuilding," *Aerospace Daily & Defense Report*, April 26, 2005b.

Capaccio, Tony, "U.S. Congress Rejects Navy's Winner-Take-All Plan for Destroyer," Bloomberg.com, May 4, 2005.

Clark, Vernon, Testimony to United States Senate Armed Services Committee, February 10, 2005.

Congressional Budget Office (CBO), *Resource Implications of the Navy's Interim Report on Shipbuilding*, April 25, 2005.

Craggs, John, Damien Bloor, Brian Tanner, and Hamish Bullen, "Methodology Used to Calculate Naval Compensated Gross Tonnage Factors," *Journal of Ship Production*, Vol. 19, No. 1, February 2003a, pp. 22–28.

———, "Reply to Thomas Lamb," *Journal of Ship Production*, Vol. 19, No. 1, February 2003b, p. 31.

Deegan, Chris, "Shipbuilding Inflation," paper presented at the 37th Annual Department of Defense Cost Analysis Symposium, February 11, 2004.

Dur, Philip A., Testimony to United States Senate Armed Services Committee's Seapower Subcommittee on Status of Shipbuilding Industrial Base, April 12, 2005.

Federation of American Scientists, "FFG 7 Program—Assignment Sheet: General Description," Assignment Sheet No. 64P7-101, undated, http://www.fas.org/man/dod-101/navy/docs/swos/eng/64P7-101.html (as of February 2006).

Fisher, Franklin M., Zvi Griliches, and Carl Kaysen, "The Costs of Automobile Model Changes Since 1949," *American Economic Review*, Vol. 55, No. 2, May 1962, pp. 259–261.

Fisher, Gene, *Cost Considerations in Systems Analysis*, Santa Monica, Calif.: RAND Corporation, R-490-ASD, 1970.

Government Accountability Office (GAO), *Could Help Minimize Cost Growth in Navy Shipbuilding Programs*, GAO-05-183, February 2005.

Hess, Ronald Wayne, Jefferson P. Marquis, John F. Schank, and Malcolm MacKinnon, *The Closing and Reuse of the Philadelphia Naval Shipyard*, Santa Monica, Calif.: RAND Corporation, MR-1364-NAVY, 2001.

Kehoe, James W., Jr., "Warship Design—Ours and Theirs," *Naval Engineers Journal*, February 1976, pp. 92–96.

Lamb, Thomas, "Discussion of 'Methodology Used to Calculate Naval Compensated Gross Tonnage Factors,'" *Journal of Ship Production*, Vol. 19, No. 1, February 2003, pp. 29–30.

Lerman, David, "Submarine Number Shrinking Gradually," *Daily Press* [Newport News, Va.], June 9, 2005.

Ma, Jason, "Newport News Chief Skeptical About Entering Commercial Ship Market," *Inside the Navy*, March 14, 2005.

Matthews, William, "Industry Officials Want Consistency in Shipbuilding," *Inside the Navy*, March 24, 2005.

Office of the Under Secretary of Defense (Comptroller), *National Defense Budget Estimates for FY 2005*, U.S. Department of Defense, March 2004. Online at http://www.dod.mil/comptroller/defbudget/fy2005/fy2005_greenbook.pdf (as of February 2006).

Office of the Deputy Under Secretary of Defense (Industrial Policy), *Global Shipbuilding Industrial Base Benchmarking Study—Part I: Major Shipyards*, U.S. Department of Defense, May 2005. Online at http://www.nsrp.org/documents/gsibbs.pdf (as of February 2006).

O'Rourke, Ronald, *Navy DD(X) and LCS Ship Acquisition Programs: Oversight Issues and Options for Congress*, CRS [Congressional Research Service] Report for Congress, updated January 25, 2005. Online at http://www.fas.org/man/crs/RL32109.pdf (as of February 2006).

Petters, C. Michael, President, Northrop Grumman Newport News, Address to the National Press Club, Washington, D.C., March 9, 2005. Online at http://www.nn.northropgrumman.com/media/speech/050309.html (as of February 2006).

Pugh, Philip, *The Cost of Sea Power: The Influence of Money on Naval Affairs from 1815 to the Present Day*, London: Conway Maritime Press, 1986.

Schank, John, Hans Pung, Gordon T. Lee, Mark V. Arena, and John Birkler, *Outsourcing and Outfitting Practices: Implications for the Ministry of Defence Shipbuilding Programmes*, Santa Monica, Calif.: RAND Corporation, MG-198-OSD, 2005.

Schank, John, Giles Smith, Brien Alkire, Mark V. Arena, John Birkler, James Chiesa, Edward Keating, and Lara Schmidt, *Modernizing the U.S. Aircraft Carrier Fleet*, Santa Monica, Calif.: RAND Corporation, MG-289-NAVY, 2005.

Sims, Philip, "Trends in Surface Warship Design, 1861–1983," *Naval Engineers Journal*, Vol. 95, No. 3, May 1983, pp. 33–49.

———, "Ships and the Sailors Inside Them," *Engineering the Total Ship Symposium*, American Society of Naval Engineers, 2004

Triplett, Jack E., "Automobiles and Hedonic Quality Measurement," *Journal of Political Economy*, Vol. 77, No. 3, May–June 1969, pp. 408–417.

Younossi, Obaid, Mark V. Arena, Richard M. Moore, Mark Lorell, Joanna Mason, and John C. Graser, *Military Jet Engine Acquisition: Technology Basics and Cost-Estimating Methodology*, Santa Monica, Calif.: RAND Corporation, MR-1596-AF, 2003.